THE PASSENGER HAS GONE DIGITAL AND MOBILE

Dedicated to My Family

The Passenger Has Gone Digital and Mobile

Accessing and Connecting Through Information and Technology

NAWAL K. TANEJA

ASHGATE

Published by
Ashgate Publishing Limited
Wey Court East
Union Road
Farnham
Surrey, GU9 7PT
England

Ashgate Publishing Company
Suite 420
101 Cherry Street
Burlington
VT 05401-4405
USA

www.ashgate.com

British Library Cataloguing in Publication Data
Taneja, Nawal K.
   The passenger has gone digital and mobile : accessing and connecting through information and technology.
   1. Airlines--Management. 2. Airlines--Technological innovations.
   I. Title
   387.7'0684-dc22

   ISBN: 978-1-4094-3502-0 (hbk)
   ISBN: 978-1-4094-3503-7 (ebk)

Library of Congress Cataloging-in-Publication Data
Taneja, Nawal K.
   The passenger has gone digital and mobile : accessing and connecting through information and technology / by Nawal K. Taneja.
      p. cm.
   Includes bibliographical references and index.
   ISBN 978-1-4094-3502-0 (hardback) -- ISBN 978-1-4094-3503-7 (ebook) 1. Airlines. 2. Airlines--Management. 3. Technological innovations--Economic aspects. 4. Customer services--Technological innovations. 5. Airlines--Cost of operation. I. Title.
   HE9776.T366 2011
   387.7068'8--dc23

2011018597

Printed and bound in Great Britain by the
MPG Books Group, UK

Contents

List of Figures

Preface

Information and technology are changing passenger behavior and passenger expectations as well as employee and shareholder expectations. Passengers now have richer and timelier information, enabled by digital, mobile, and social technology. They are now empowered, for example, by not only price-service comparison websites, but also user-centric rating websites. Passengers now want to be in control and expect airlines to become solution providers and aggregators of value to provide much more personalized services and better travel experiences across the travel cycle. Airline employees expect to be given information- and technology-enabled tools to do their jobs to meet passenger expectations by accessing and acting upon real-time information. Shareholders expect returns that are reasonable and relatively stable through business cycles. Airlines can now use insightful information and enabling technologies to target more intelligent segments of the passenger base and serve each selected segment with profitable price-service options. These passenger, employee, and shareholder expectations can be met by airline managements through further innovation of business models that start with an increasing focus on customer centricity (from operation and product centricity), followed by a better understanding of newer customer segments—maturing sub-segments in North America and Europe, growing middle classes in Asia and Latin America, and the divertible segments from trains and buses in countries such as India and Mexico. Next, the evolving business models need to focus more on different price-service option packages for strategically-selected segments while minimizing the complexity and costs of producing and delivering cost-effective services for these segments.

At the beginning of the first decade, the first book in this series[1] (*Driving Airline Business Strategies through Emerging Technology*) was devoted to the then new technology, anticipated to affect both buyers and sellers of airline services. Even that technology, encompassing all areas (software, hardware, and applications), was capable of enabling airlines to access vast quantities of data from disparate sources and to make sense of this data, for both passengers and airlines, and to benefit passengers and airlines. Technology was available, for example, to enable airlines to gather, synthesize, and mine relevant data to segment their customer base differently not only to provide much more cost effective value propositions to different sub-segments, but also to have two-way interactions with their passengers to provide solutions to the problems of their segmented passengers. The data at that time was collected from traditional sources, such as passenger surveys, airline call centers, and frequent flier programs. A few airlines did capitalize on this emerging technology, even though their initiatives were limited relative to some non-airline companies, taking advantage of best global practices. Within the airline industry, when the Internet became a widespread technology and more and more households got connected, it was the low cost airline sector that began to deploy this component of the new technology to sell tickets through airline websites, thus reducing drastically the distribution cost. Traditional legacy airlines, in general, were a little slow to capitalize on the evolving technology for at least four reasons. First, they had seen the Internet stock bubble burst and the skeptics wondered if this new wave of technology was simply a fad. Second, many found it somewhat difficult to change the internal corporate culture (relating, for example, to the inward-thinking and the historically-based practices and systems). Third, many airline information and technology officers were not able to "develop and sell" persuading business cases due, in some cases to the lack of clear technology strategies, and in other cases to the lack of alignment between information and business strategies. Fourth, they had *legacy distribution models* in place, namely, huge call centers and reliance on travel agents with elaborate commissions. It takes time and money to implement these changes.

A decade later, we have new sources of information, exemplified by Web traffic, Web content, Web behavior, Web analytics, and communications through social media. At the same time, technology advancements have moved to a higher level (search engines, mobile devices, analytics, applications, and social networks). Within this framework, a fundamental change in technology relates to the proliferation of "the third screen," as discussed by author Chuck Martin in his book, *The Third Screen*. The third screen, the mobile device, (the other two being, of course, the TV and the PC), not only provides buyers with a means to consume information on the go, but it also provides sellers with a challenge and an opportunity to engage with the buyers.[2]

There is also now a plausible explanation for the bursting of the Internet stock bubble. The IPOs of many newly-formed Internet-based companies had way overvalued stock, given the lack of their business fundamentals and wild assumptions. For example, profits were based on not the sale of products and services, but on the number of "eyeballs." On the side of unrealistic assumptions, one was that retail stores would become obsolete. Such is not likely to be the case with the new-generation technology. For instance, relating to air travel, it is not assumed that new-generation technology will make air travel obsolete and that business travelers would rely on video conferences to conduct their business while leisure travelers would become satisfied with taking "virtual" trips. Consequently, new-generation information and technology can be used to bring innovation to airline business models while keeping intact the airline business fundamentals.

This book, the seventh in a series with Ashgate, explores potential game-changing opportunities available to the global airline industry through the generation and analyses of new-generation information and the deployment of enabling technology. It provides examples of how technology can now enable the alignment of information and business strategies of different categories of airlines to become truly customer centric at producing and delivering cost-effective personalized services based more on information than the physical product attributes. The large conventional network carriers can deploy, for example, information and technology to mobilize their inherent strengths (huge global networks, wide alliances, diversified fleets, and so on) to compete more effectively with the LCCs. The LCCs can, in turn, use their own inherent strengths (agility, light and flexible structures) to adapt swiftly to the ever-changing environment. Even more importantly, technology can now enable the two sectors to cooperate for mutual benefit.

As with other books in this series, this one also presents numerous mini case studies of best global practices that leverage information and technology. The main audience of this series continues to be senior-level practitioners of different generations of airlines worldwide as well as related businesses. As with the previous books the approach continues to be to provide impartial, candid, and pragmatic analyses of the role that technology can play in enabling new-generation information to bring innovation to different airline business models.

Notes

1   The first, *Driving Airline Business Strategies through Emerging Technology* showed that in the rapidly evolving airline industry, emerging technologies could indeed play an increasingly critical role in the delivery of real and perceived customer value. The second, *Airline Survival Kit: Breaking Out of the Zero Profit Game*, wrestled with the precipitous decline in the profitability of the industry and discussed some strategies for dealing with the heavy burden of excessive complexity incorporated within the operations of legacy airlines. Having realized that the industry and the environment were experiencing step changes, the third, *Simpli-Flying: Optimizing the Airline Business Model*, drilled deeper into the discussion on restructuring of markets and the critical need for strategies to adjust to the new aviation realities. The central theme of the fourth, *Fasten Your Seatbelt: The Passenger is Flying the Plane* was that core customers—not airline management—are beginning to seize control of the direction of the industry. The fifth, *Flying Ahead of the Airplane*, analyzed global trends and provided some thought-provoking scenarios to help airline executives adjust and adapt to a chaotic world. The sixth, *Looking Beyond the Runway: Airlines Innovating with Best Practices while Facing Realities*, argued that the global airline industry is now facing a "new normal" and, as such, airlines must now go beyond their short-term circumstantial strategies and renew their ageing business model.

2   Chuck Martin, *The Third Screen: Marketing to Your Customers in a World Gone Mobile* (Boston, MA: Nicholas Brealey Publishing, 2011), p. xvi.

Foreword

Akbar Al Baker
Chief Executive Officer
Qatar Airways

Nawal Taneja has put together a persuasive argument on how to become a customer centric airline. There are many game changing concepts and ideas in this book which can provide the manual for CEOs, CIOs and others to recreate their company.

At Qatar Airways we are especially receptive to the fact that customer centricity is fundamental to be a successful airline. We also believe that it fits in line with our 5 star customer service and product. Caring about the customer not only requires hard and soft products, but also the right attitude and steadfastness to see customer centricity through to the end. Technology changes are providing the ability for us to adapt and become even more engaged with our customers at every touch point in the journey. The winners will need to move fast and learn from the best in class not only from our industry but others. That is why you see Qatar Airways designing products not just based on airline trends but also on broader hospitality trends and customer needs. This is why, for example, our Premium lounge in Doha feels more like a hotel than an airport lounge.

There are many exciting opportunities and Qatar Airways welcomes in an era where customer expectations are now heightened by companies like Apple, Google, and the social media. We are a young and vibrant airline and feel well placed to emerge as a technology and customer centric leader and winner in the connected and the increasingly demanding customer base. Customers are expecting airlines to reinvent their long-haul travel experience. They are expecting airlines to install in-flight connectivity as well as mobile and Wi-Fi. Qatar is committed to not only to meet, but to exceed customers' expectations as evidenced by the quality of our long-haul services.

Nawal has effortlessly put the language of the internet, social media, etc. into airline terms. He has shown the path, the hurdles and the opportunities. However, every airline remains unique. We truly believe that a fast growing airline has even more opportunities since we are always changing and open to change. Our organizational structure is fluid enough to cope with the integration of technologies and the breaking down of silos and barriers. It is an exciting time to run an airline. It is exciting to be judged by our customers in real time on a daily basis. However, it is even more fulfilling to respond to them in real time and offer them the best products in our class.

Qatar Airways is equally content that we are already deploying some of the technology mentioned in this book. We are already measuring and managing on the new set of metrics from a customer's perspective. We are open to the new mindset. Another advantage Qatar Airways has is that this openness is receptive to all of our labor groups. We also have the benefit of long term investment vision and a solid approach to return on investment through the fulfillment of customer and employee needs.

What makes a truly great company is one where the employees come in every day and ask:

"How can I make a difference for our customers?"

"What can I do new today?"

"What can I improve?

This attitude is at the core of what Mr. Taneja is speaking about. It is a great challenge and we accept it.

Doha, Qatar
July 2011

Foreword

Eric R. Conrad
IBM Travel & Transportation Global Industry Leader
Global Business Services

One hundred years ago IBM first began developing machines to count, sort, and analyze the business world. Many of these machines transformed the way business was done. And in some cases, these machines quite literally changed the world.

One such transformation occurred in 1964, when IBM and American Airlines launched the airline industry's very first online reservation system, SABRE. The system connected ticket offices across American's network with a central computer that provided real-time access to their inventory. SABRE replaced a labor intensive process of hand-written cards with an interconnected, automated process that enabled airlines to sell and manage bookings worldwide, instantaneously, and up to a full year in advance.

We all know the implications of this great leap forward: airline marketing and sales processes were changed forever. Customer service was greatly improved and tremendous inefficiencies were driven out of airline operations, allowing the industry to grow.

Today, the airline industry stands on the precipice of another potential transformation. The immense growth in the volume of available data—from the millions of transactions each day to online interactions with customers through social media and mobile devices—is creating both opportunities and challenges for travel providers.

In Nawal Taneja's book *"The Passenger Has Gone Digital and Mobile: Accessing and Connecting through Information and Technology,"* he discusses many of these challenges and shares his insight into possible strategies and responses. Even to the most prescient leaders in the industry, this book will provoke new and valuable thoughts as they develop their own transformation agenda.

At IBM, we believe the most successful airlines will transform by taking advantage of computing power and analytical models to gain understanding from their seas of data. The core challenge is to develop context around each piece of data to derive deep insight about the intentions and specific needs of each customer. In this way, the industry will engage in more personalized interactions with its customers and provide offers or services that are aligned with an individual's needs or preferences. Those that do not rise to this challenge are very likely to be left behind.

Let's look at a few examples:

- *Marketing and sales*: The airline that can detect the personalized needs of a customer who interacts with them over its website—through historical data, how customers use the site, behavioral modeling, and contextual factors like time of year, O&D, weather or events—is in an ideal opportunity to create an offer that includes more than just the flight itself and builds brand loyalty not seen today. This could include ancillary services and travel-related offerings from transportation to hospitality to entertainment services that are entirely unique to an individual customer.
- *Irregular operations*: The airline that understands who and where their passenger is when there is aircraft delay and recognizes her travel plans and patterns, past behavior and preferences during a delay, and the value of that traveler to the airline is in the perfect position to proactively rebook the traveler over her preferred communication mode and offer services and information that uniquely fits her needs at that precise moment, whether that means lounge passes, coupons, navigational services, onward travel arrangements (e.g., rail), and more.
- *Core operations*: The airline that connects and analyzes the wide sources of structured and unstructured data, both internally and externally (e.g., social media), to understand emerging operational issues can better respond to those issues in real time combining insights into the actions of its competitors and partners with an awareness of the location, availability and performance of all of its assets and people. They can not only avoid becoming the headline story on

the nightly news, but in doing so, will likely reinforce the positives of its brand and, likely, market share.

The facts are clear: the proliferation of devices (e.g., mobile, tablets, in-vehicle computers, home entertainment), communications (e.g., bluetooth, near field communication, RFID, WIFI), and information (e.g., GPS, social networking, advanced search engine technologies, and software as a service) has created an unprecedented opportunity to deliver services to customers in new and exciting ways—from marketing to selling to delivering travel. From services that exist today (e.g., ordering food, selecting entertainment) to services that are just being dreamed of (e.g., multi-modal seamless travel). Imagine being able to guide a passenger from the alarm she needs to wake up and get to the flight on time, to the drive to the airport, the navigation through the security processes, directions to the lounge, gate and onward travel and then baggage collection, transportation, and additional connections and services to fulfill her journey.

The age old pressures of reducing costs, building new sources of revenue, improving customer service and weathering the inevitable externalities that have impacted airlines through the years will never go away. But to them will be added the exciting challenge of driving more unique and valued services, regardless of the nature of the airline. Our instrumented, interconnected, and increasingly intelligent world offers great potential to airlines in their quest to differentiate their services. Deliberate and thoughtful analysis of the available data—to determine context, insight, intent and offering—will provide a clear path to success in the years to come.

In 2011, IBM's Watson computing system won a historic match of *Jeopardy!* against two of the game show's all-time champions. This match not only demonstrated Watson's natural language Deep Question and Answer (QA) capability, but also the underlying data management and deep analytics. Consider that Watson analyzed over 200 million pages of structured and unstructured data and came up with the best possible answer in just 2–3 seconds to answer a question.

The importance of this match is not the accomplishment of winning *Jeopardy!*, but where this effort will take businesses.

For airlines, Watson represents the promise of what the future might hold to help solve some of the toughest issues from fuel optimization to crew scheduling and resource deployment to operational recovery, contact center automation and truly unique traveler personalization.

The innovative capabilities that powered some of IBM's early machines that made industry transformation possible are still at the heart of our company. And we believe that the future of the airline industry will be driven by close partnerships that endeavor to understand and solve the industry's most urgent challenges. Nawal's timely and thoughtful analysis should initiate a productive dialogue across the industry on how to address some of these challenges, and, with sustained effort and a little luck, result in tangible progress toward innovative solutions that will set new standards of customer satisfaction and corporate profitability.

Singapore, Singapore
July 2011

Foreword

Constantino de Oliveira Jr.
Chief Executive Officer
GOL Linhas Aéreas Intelgentes

In his new book Nawal Taneja brought to the spot light the enormous opportunities that the airline industry has ahead by gathering concepts and tools from Information Technology. These ideas are particularly applicable to low-cost carriers (LCCs) whose business model is evolving from the operating-centric approach to the passenger-centric approach not abandoning, however, the revolutionary conquests they have made since the 1970s. In the case of GOL we started to provide attractive price-service options beginning in January of 2001 and stimulated demand.

On the other hand, this book addresses some key questions of full-service airlines relating to how they can provide differentiated services while not abandoning the cost-efficiency perspective by the application of Information Technology concepts and resources.

Experience has shown that low-cost and full-service airlines have begun to converge their business models and Information Technology has been an important element to facilitate this convergence. The time of the classical duality of cost leadership versus differentiated strategic positioning is about to be over. Thus, in the classical LCC business model interline agreements, code-sharing, medium or long-haul flights, and frequent-flyer programs were unthinkable and these airlines were typically "stand-alone carriers." These supplementary services were attributes of the legacy carriers and now they are beginning to be integrated in the new generation LCC business model. Southwest and GOL are two good examples.

Over time LCCs have expanded very much their market share. However, they cannot focus exclusively on the price-sensitive demand anymore. In order to attract new segments of clients to their flights they had to abandon the "stand-alone carrier" concept and they had to become connected to the "legacy-airline world." In parallel, the legacy carriers had to become more cost-competitive as a pre-condition to survive. They had to understand that nowadays air travel is a commodity and thus price is one of the most salient attributes considered by clients in their process of choice of a specific flight. The response to this apparent paradox for the airlines—need to improve services and to reduce costs at a same time—is Information Technology and it is a common response for LCCs and legacy airlines as well.

I remember when we started GOL 10 years ago our lemma was "Lower fares, higher load-factors, strong profit growth, lower costs and lower fares." We used to call this say "the virtuous cycle of profitability." Our strategy was to invest in modern and standardized fleet, Information Technology and, most important, highly motivated staff. At that time, as today, to invest in Information Technology meant to make easier the passenger's life by adopting, for example, a ticket-less booking system and all its corresponding implications. The strategy has worked and GOL has become one of the fastest growing and one of the most profitable airlines in the world at the same time.

Passengers, as well as clients of many industries, have evolved and their expectations are much broader. Fortunately, technological tools evolved in the same way. The airline industry now has the opportunity to lead this transformation process as it made in the past in other innovation movements, such as the globalization in the 1990s.

My perspective is that in the years ahead the airlines that are going to survive in this turbulent industry are those that have elected to develop and implement the passenger-centric approach as the main objective and Information Technology as the most important means to fulfill the passenger's growing needs. In this context, to understand that *"the passenger has gone digital and mobile,"* is the first step to be accomplished by the airline executive who intends to lead his or her airline to become succesful.

I am absolutely convinced that the digital and mobile capabilities and strategies of an airline are the key elements that will help an airline to move up to a higher level of performance from an average performance. This book is an important contribution to help executives to place his or her airline in the first group.

São Paulo. Brazil
July 2011

Foreword

Tewolde GebreMariam
Chief Executive Officer
Ethiopian Airlines

The success of Ethiopian Airlines Vision 2010 was based on the rapid and profitable expansion of its route network (domestic, regional and international), by both increasing frequencies to existing destinations and adding new ones to its network. In so doing it has over the last five years, 2005–2010, trebled its fleet, grew constantly its number of passengers despite the great recession and also expanded its cargo and other aviation-related businesses which are important revenue contributors to its strategic objectives. It has also invested in its people through continuous training dispensed by the Airline Aviation Academy and adopted new technologies to improve its various internal processes across the airline and associated supporting business units encompassing, amongst others, ground handling and catering.

Motivated by this successful achievement of the 2010 Vision the airline has set itself an even more ambitious objective in its Vision 2025 which, besides modernizing and expanding the fleet, includes further network expansion either through direct operations or through its code-sharing and partnership in the STAR alliance. Having won the battle of product harmonization as a result of a new young fleet, which will also include the newly born and highly technologically sophisticated aircraft, the Boeing 787 Dreamliner, the airline today, more than ever, puts customer centricity, efficiency and cost effectiveness at the centre of its future development.

Ethiopian Airlines, being fully conscious of the critical role of information technology and systems for achieving its set goals, has started since last year a major IT revamping which, besides enhancing the efficiency of its internal business processes and ensuring integration across value chain, also aims at improving proactive decision making at all levels through adoption of business intelligence. With customer relationship management (CRM) tools the airline aims to have a yet better understanding of the customer in view of enticing loyalty and also to bring the airline even more closer to customers through personalized services and new direct distribution and interaction channels such as the Internet, mobile computing and social media.

Nawal's book entitled *"The Passenger Has Gone Digital and Mobile"* is most timely as it directly addresses the major concerns we are presently facing in this information age where business speed is of the essence, customers are getting even more sophisticated and demanding and they want to have timely information at their fingertips at all times, competition becoming fiercer and stiffer as all stakeholders aim at either improving their market dominance or capture a bigger share in the travel value chain. Omnipresence and responsiveness across all forms of the media become vital as customers are only two computer clicks away from your nearest competitor and thus can easily switch their business.

With this in mind and to support its Vision 2025 Ethiopian Airlines built its IT strategy on five main technological solutions, namely,

- a scalable, secure, performing and adaptive infrastructure encompassing LAN, WAN and redundant data-centres,
- a portfolio of applications supporting and enhancing the core airline business processes such as reservation and ticketing, departure control, MRO, flight operations including crew optimization and cargo systems,
- a portfolio of applications supporting and enhancing the corporate functions finance, logistics, human capital management, employee self-service, collaboration and productivity tools,
- a suite of decision making tools around business intelligence and revenue management, and
- a suite of approach encompassing marketing solutions, E commerce, CRM, mobile computing, and social networking with the aim to enhance the overall customer experience and thus entice loyalty.

Besides the technological solutions the Ethiopian is adopting business process reengineering as per recommended airline best practices coupled with efficient project management and change management approaches to ensure user acceptance of new work practices. Continuous training for both end users and technology staff is paramount for successful adoption.

As we succeed in removing the historical organizational silos culture and promote better end-to- end process integration within the company, Ethiopian Airlines is now aiming at strengthening integration of business processes beyond its own boundaries and transcending to partners. Indeed in the areas of procurement the airline is looking for backward process integration with key suppliers, aircraft manufacturers and spare-part providers. In its new alliance in the Star community, information technology is poised to play a key role in interacting with partner airlines, so as to make the customer experience seamless. Information technology is thus becoming ubiquitous and fully integrated in Ethiopian Airlines' daily activity.

We have judiciously contained our overall IT expenditure by adopting the strategy of Acquire and Manage whilst favouring Suite Adoption wherever possible to reduce overall integration costs. We are thus using the SAP suite for Finance HR and Logistics, Sabre Suite for our Passenger Management Systems and Flight Operations, and the SITA Solution for Cargo and Maintenix for all our MRO business processes.

Our customer-facing applications are being revamped to induce customers to interact with us online. Besides being informative the website is becoming more transaction oriented to support both customer service and sales. Through the Web 2.0 approach Ethiopian is encouraging customer feedback and community portal. In line with revenue management policies the booking engine, which is being reworked, will provide more appealing choices through calendar shopping, wider coverage by including connecting flights with partner airlines and package solution, including air and non-air products. Enrolling for frequent flyer, tracking of cargo shipment, and latest updates of flights schedules are a few customer service functions. The airline is also building links with other online travel portals like Expedia and is following travel blogs or other relevant social networks, including multimedia broadcast through YouTube, and will add content whenever it makes business sense. Through SMS alerts the airline is making its first start to mobile computing which promises to be a long interesting journey as booking, paying for travel solutions and interactions through this medium become most popular.

However, as the CEO of Ethiopian Airlines, I have a yet another major challenge as my customers come from over the whole planet, United States, Europe, Asia and our home market Africa and there is no homogeneity in the infrastructure available, in the level of knowledge and technology adoption of the people and businesses. For instance credit cards, which are critical for e-commerce, do not have the same penetration in Africa as compared to the rest of the world. Bandwidth and reliable telecommunications are still issues in various part of Africa. I am nevertheless optimist that this digital divide will soon be filled as various national initiatives are pushing for infrastructure modernization and adoption of the state-of-the-art technology, enabling leapfrogging, as we do not have legacy constraints. Whilst not having been fully abreast with the Internet development, early and intensive adoption of mobile computing can help the lagging markets to catch up.

Whilst I am all in favour for these new emerging technologies as I am convinced that it will help me to reduce my cost of operations, improve my customer's experience and enhance revenue, while optimizing the total cost of ownership, I cannot disregard their pitfalls, when wrongly exploited, as testified by the various fraud stories, hacking and security breaches, and dis-reputations caused by alien and unfriendly entities. Such unreasonable behaviour can washout all efforts of loyalty building and brand positioning, built over years by such social networks that unfairly ill-talk and bad-mouth a company or its services.

Ethiopian Airlines is thus vigilant and optimistically cautious in the adoption of all these unavoidable technologies, and will ensure continuous appraisal and communication of its services to stay at the helm of the industry and in the heart of its existing and prospective customers.

Information technology has always reshaped our industry from the early age and adopters have always gained competitiveness positions. In the early sixties the emergence of automated reservation systems reshaped the interaction between airline branches, travel agencies and the airline, thus giving substantial edge to their owners. The emergence of global distribution systems in the late eighties further reshaped the industry and cost of distribution increased drastically so much so that airlines were looking for disintermediation. The solution came with the Internet in the late nineties when airlines had their own websites and booking engines. In the same breath online travel agencies were born. All these developments were pushed by technology bringing self-sufficiency to the customer.

In this decade we have moved even further where the customer is king and in control through social networks. The pace of change through technology is getting even faster and airlines must be smart to choose the functionalities and devices that are going to make a difference to their business. Businesses will fail if they are not agile in responding promptly and flexibly to the demanding and sophisticated customers who expect the same sort of service they are getting from the retail or banking industry. Indeed, information technology has now moved from being a business enabler to be today an essential business driver and source of achieving competitiveness. I view the future with thrill but also with some reservation due to business uncertainty in this new era of fast technological change.

I feel most privileged to author a foreword of the latest book of Mr. Nawal Taneja, a well- known figure in the Air Transport Industry. In this new book, entitled *"The Passenger Has Gone Digital and Mobile,"* he brings us to even newer paradigms which will reshape customer behaviour and the process through which airlines are going to interact in the future. He is consistent with the idea that passengers are in control and airlines must innovate and evolve their business model towards more customer centricity whilst harnessing the full power of emerging technologies to deliver a flawless passenger experience. The mini case studies give good insight on best global practices leveraging information and technology.

Upon this I conclude by asking you to *Fasten Your Seat Belt* (the title of one of his earlier books) and enjoy an easy to read book free of Geeks jargons and full of insightful ideas for our business—the Airline Business.

Addis Ababa, Ethiopia
July 2011

Foreword

Vice President, Principal Analyst, Airline and Travel Research
Forrester Research, inc.

If you're not customer-centric, do you *really* have a viable business?

We are doing business at a time that we at Forrester Research call "The Age of the Customer." Today's consumers are hyper-connected, empowered, and disrupting nearly every industry, including airlines, media, retail—even governments.

Consider these facts:

- In the US, where on average 22 percent of adults own a smartphone, 35 percent of leisure travelers and 55 percent of business travelers carry one of these devices. Smartphone adoption is expected to grow by more than 60 percent in the next five years. Among these smartphone-toting travelers, half of leisure travelers, and almost 3 in 5 business travelers, have already downloaded at least one travel-related app.
- In Europe, in 2010 72 percent of online leisure travelers participated in social media. That rises to 77 percent in the US (as of Q1 2011) and climbs further in Asia, where 4 in 5 online adults in Japan, and 83 percent of online adults in China, participate in social media. Almost half—48 percent—of US online travelers say that traveler-written online ratings or reviews are important to them, dwarfing the number (27 percent) who value the advice they get from a professional travel agent.
- Intercontinental Hotels Group reported that mobile is its fastest growing revenue channel. In less than a year, the global hotel giant saw mobile-generated revenue soar from $1 million a month to $10 million a month. Sixty percent of those mobile bookings are made within 24 hours or less of a guest checking in. Priceline, one of the largest online

travel companies, reports that 35% of the people who use its mobile application to book a hotel room do so within one mile of the hotel at which they stay.

As if that's not enough, airline loyalty programs, which celebrated their 30th anniversary in 2011, may not necessarily be as successful as think they are. Some of these programs have membership numbers that exceed the population of major countries. And while programs like Mileage Plus, Flying Blue, and Marco Polo may generate substantial revenue for their carriers through the sale of miles, kilometers, or points to third parties, few travelers consider themselves to be loyal to a particular airline. In the US (the only market where this information was available at press time), just 36 percent of airline passengers felt they were loyal to any single airline.

As a result of smartphones, social media, and an accelerating pace of technology development, we are doing business in the era of now. Smartphones enable what we call the "story arc" of opportunities. Smartphones and the story arc make it increasingly practical, and essential, for airlines to parse a journey into a series of micro-journeys, from which various marketing, sales, and service messages can flow—both from carrier to traveler, and traveler to carrier. Within three years—and possibly two— we at Forrester believe that smartphones (along with traditional websites) will become the two most important digital touchpoints airlines use.

Social media allows airlines to have true, meaningful engagement with travelers, employees, and other stakeholders. Social media also means these same groups will have their own conversations about you – in fact, they're doing so right now. Airlines like KLM, Virgin America, Delta, AirAsia, Southwest, and JetBlue are leading the industry through effective uses of social networks like Twitter and Facebook. They, and others, use social networks mostly for marketing and customer service and support, but the first blossoms of social commerce have emerged. JetBlue has a dedicated Twitter handle for promotional fares (@ JetBlueCheeps), and Delta Air Lines has a booking engine on its US Facebook fan pages. KLM monitors its fans' social networking updates and surprises randomly chosen travelers with small gifts as they fly through the carrier's Amsterdam hub. One in 5 US online travelers will consider planning or booking travel through a social networking site within the next year.

As discussed throughout this book the pace of technology change is indeed rapid. As of July 2011, Apple iPads alone are estimated to account for one percent of global Web traffic. Remember, this is a device that was introduced less than three years ago -- during the most severe recession since the 1930s. Google is working on the integration of ITA Software, has developed a smartphone equipped with a built-in Near Field Communications (NFC) chip, and has launched a virtual wallet. Within a year, the first blossoms of Google's travel distribution strategy should become clear. Within three years, we expect Google will be among an airline's five most important distribution channels. Is that good or bad? Only time will tell.

In the Age of the Customer, and the era of now, airlines need to be honest with themselves and with their customers. Don't promise a filet mignon experience if what your strategy calls for is hamburger. The customer experience matters, but it needs to be true to your brand and its value proposition, a point also made by Nawal in this book. Key to forming and supporting these experiences are technologies like smartphones and social media. But never has the saying "live by the sword, die by the sword" been more meaningful than it is today. Your successes and failures will be tweeted and shared — perhaps even using the in-flight Wi-Fi your carrier offers as an ancillary product to your travelers.

San Francisco, California, USA
July 2011

Foreword

Peter Hartman
Chief Executive Officer
KLM Royal Dutch Airlines

"From now on, the customer drives the customer experience"

Only a few years ago, customers just had to deal with our complicated processes and back office labyrinths. Check-in was Ground Services' focus, the booking tool belonged to Sales and loyalty was Marketing territory. Nowadays, the lines between sales, services and marketing are blurring. Silo-thinking is no longer accepted. Customers want to choose a single point of entry, and they have become agile in alternating swiftly between different functions.

We took on this challenge in April 2010, when a volcanic ash cloud from Iceland drifted across Western Europe. Safety precautions grounded our flights for four days, triggering a deluge of tweets and Facebook posts from stranded customers asking for help. Our existing communication channels had to operate under extreme circumstances. That was when we decided to deploy our nascent social media activities for customer service purposes. We informed passengers through Facebook and Twitter. Customers encouraged us to use these platforms for Q&A's, for servicing and even for rebooking their flights. Over 100 volunteers from all over the company worked 4 hour shifts to assist our customers 24/7. And it hit the bulls eye! By the time our aircraft took the skies again, we had established an exceptional and effective fast track.

This experience changed our business drastically. Soon after, the "KLM Social Media Hub" was set up. At this heart of our social media channels, employees from across the company, including e-commerce, corporate communications, customer care, in-flight services, IT and marketing, joined forces. Today, 23 people are dedicated to handling all incoming requests via social media around the clock. The promise: response within the hour, and a solution to any kind of problem within 24 hours. Social Media have become a "mission control" for customer complaints and praise, a measuring point for sentiment, as well as a new place to build our brand.

Only a few months later in the same year, we faced a "snow crisis" which hit all major hubs in Europe. We immediately experienced the effect of our previous actions on our customers' expectations. They now looked for the same services through social media as during the ash cloud period. The lesson learned: once you start, you cannot go back.

To reinforce the customer experience there should be focus on all customer touch points: the KLM.com website, the self service machines at the airport, on board, KLM.com for mobile and now also social media. Customers now not only decide what they want to buy and which information to get, but also when, how and where. This is quite a challenge. When providing information or selling a product, we should be consistent and relevant throughout all touch points.

As Nawal Taneja states, customer centricity is key. And I also agree with him that adopting the latest technology only is not enough. The changed customer behaviour enabled by mobile developments, new acquisition channels and applications is indeed a game changer for our business model. But I believe the reward is clear: quicker decisions, happier customers and stronger loyalty. And we all know that loyal customers will return.

Amstelveen, The Netherlands
July 2011

Foreword

Dr. Christoph Klingenberg
Senior Vice President
Information Management and Chief Information Officer
Lufthansa German Airlines

This book gives a comprehensive overview of the massive changes the airline industry faces through new media, new devices and the new generation of travelers. Many of these developments are already underway, challenging traditional paradigms, processes and organizations. As the CIO of an "old world" airline from "old Europe" I am witnessing changing customer behavior and the need for a more agile business setup. IT can certainly enable business with the means to cope with these challenges. However, as Taneja points out, it needs strategic planning and adequate resource allocation to make a difference.

One of the topics in this book I like most is customer segmentation. Not only can a smart segmentation serve as a means to provide better customer service, it also can help generate more profit. Except for corporate rates airlines are almost completely ignorant of the person who is looking for flight (and price) availability. In some cases even corporate travel managers look for spot prices on the intended travel days that sometimes may be lower than the then contracted corporate rates. Once airlines can get the GDSs to collect this information they can incorporate it in their revenue management systems, enabling the customization of prices based on the willingness to pay and based on the actual product requested (flexibility, comfort). There are some trials being currently conducted in the market in the area of closed user groups, for example offering cheap tickets to students. However, this method has to be much more refined to work also for customers who are not willing to register for a certain group (and thus remain flexible to reflect their changing travel needs when on business or leisure). This will only work, however, if the

GDSs transform themselves from a mere distribution system to a customer centric and value added channel.

The book points out that the new technology makes much more information available to the passenger, and that this offers huge opportunities to engage them for tasks that traditionally were taken care of by the airline. Since self services like check-in and mobile boarding passes have almost become industry standard, it is now time to enable the passenger to change his booking in case of irregularities through his or her mobile device—thus becoming independent of airline ticket counters and saving hours of waiting time and much annoyance. I personally, however, do not agree with the proposed strategy of equipping large number of staff with mobile devices to ramble around at the airport and serve stranded customers. While this may help on board (as many other airlines, Lufthansa currently equips their pursers with iPads), on the ground it is very challenging to recruit enough standby staff during heavy weather situations. I think it is better to hand out mobile devices to passengers and to simplify the processes so that most customers can handle their rebooking needs themselves.

Is it really necessary that at every touch point all the information about every customer must be available? Data protection issues aside, if a passenger experienced a disservice, then a separate process within the airline must be initiated to remedy it. I do not believe in telling this customer at his next check-in or during the next inflight service, while handing over the coffee, how sorry the airline is to have inflicted this disservice upon him.

Taneja points out that many of the investments necessary to compete in this arena are largely of experimental character and cannot be calculated and justified as most other investments in the airline industry. This is due to the fact that the technology is still in its early stages and that customer satisfaction cannot readily be measured in hard dollars. So, at Lufthansa these investments are treated like investments into new seats or lounges.

One of the caveats that Taneja alludes to in Chapter 4 is that in spite of all new technology developments airlines must not loose focus on delivering their basic product right. While new gadgets can shorten perceived waiting times and mobile devices can empower customers, the airline has to deliver on-time performance with reliable and friendly service.

Whatever the next couple of years will bring, the new media will increase transparency and customer choice. Airlines with a solid offering, a good cost position and proper customer orientation will embrace this change, since it will strengthen their market positions. And this book makes fun reading giving you the flavor of the future.

Frankfurt, Germany
April 2011

Foreword

Tan Sri Dr. Munir Majid
Chairman
Malaysia Airlines

Full service airlines must transform themselves if they wish to succeed and not be driven into oblivion. Fundamental changes to business models are necessary, even if there is undoubted need for change as well to business processes that bring value to the customer. Hence, technology is an important enabler.

However while IT is a strong tool and enabler which can drive strong benefits, it would be foolhardy to think of it as an end in itself. Unless the software is internally developed and proprietary, the first-mover advantages from its adaptation do not last long enough before the enablers are commoditized and we are back to square one insofar as the competitive terrain is concerned.

Which major industry in the world has as many as 230 companies, just going by membership of IATA and not counting the formidable low cost carriers, competing for the same piece of the action? There is such a waste of resources in terms of duplication of costs.

No wonder profit margins are so tight and are so easily wiped out when cost items such as jet fuel rise. So what are we in business for?

For the world's first truly global business, the full service airline industry is the least global in mind. It has become too dependent on protection and closed skies, riding on a nationalistic credo out of place in a globalized and competitive world.

The industry must embrace change more fundamental than the adaptation of enabling technology which is necessary but not sufficient.

Internal strengthening alone would not work. A full-scale reappraisal, a strategic clean slate transformation of organization and business activities, was necessary. Full service carrier companies must have their low cost arm as this is where business growth is greatest. They must develop and diversify ancillary activities to boost revenue and profits, such as in maintenance, repair and overhaul activities, the cargo business and training.

Most of all, there must be consolidation of the industry across boundaries and businesses. There have to be big time mergers and acquisitions, significant joint ventures. The airline alliances are only half-way houses which stopped far too short of their full promise. We must move fast on consolidation. Otherwise many, many airlines will be driven into oblivion.

The changing dynamics of the airline industry and the convergence of technology are changing the game so rapidly. The convergence of technology makes innovation the driving force to enable us to stay in the game or change it for survival.

As discussed in this book, Malaysia Airlines had scored many firsts with its MHmobile for booking, check-in, flight status updates, timetables and baggage tracing, use of iPad for ticketing known as MHkiosk and launch of Augmented Reality using the GPS on the iPhone. The airline has also taken the social media to new heights by allowing passengers to book their flights and to check in through the Facebook. Passengers can integrate their trip details into the social graph, identifying friends who may be on the same flight or destination.

The chief information officer should play a bigger and more interactive role, both with the chief executive and with colleagues across the business organization. He or she must align and understand the business view. We need to leverage on technology to generate insightful intelligence and use that intelligence to expand our revenue potential and transform the value chain. The author discusses these points in detail.

Incremental changes to our business are no longer sufficient for us to deal with the challenges and opportunities facing the airline industry. Change needs to be transformational – enabled by unconventional thinking and the deployment of game changing technology to produce compelling and personalized value propositions.

Kuala Lumpur, Malaysia
June 2011

Foreword

David Neeleman
Chairman
Azul Brazilian Airlines

Having founded successful airlines, I learned a little bit about what differentiates a good company from an outstanding one. I follow three basic steps: 1) take good care of your employees so in turn they can look after your business, 2) be impeccable when executing/delivering your services or products, and 3) treat your clients even better when things go astray. To achieve these three foundations of my companies, I have always relied in technology. It helps you build a better, more efficient product. Moreover, it frees up your time so you can concentrate on introducing improvement. That is why I introduced ticketless travel and related technology at JetBlue. That is also why I was a pioneer in deploying live TV in the cabin as well as other novel ideas such as the deployment of electronic flight bags in the cockpits. And finally, that is why I started letting our call center employees work from their homes instead of having to commute to the office everyday.

I really felt honored to be asked by Mr. Taneja to write a foreword for his latest book. In this book, the reader will find fundamental lessons on how to better deploy information, business intelligence, and enabling technology to improve the success of a business. This is exactly what Azul has been doing in Brazil.

We clearly identified an untapped market segment, namely, people who typically traveled on interstate bus services, very often taking days on end to cross the vast expanses of the Brazilian territory. In fact, a sizeable chunk of the Brazilian population had never hopped on an airliner.

We developed products for this segment of the market and are enjoying considerable success. For example, we offer low, unrestricted fares with a 30-day advance purchase. While such a fare concept might not have been new in other parts of the world, it was not very common in Brazil. Developing a low fare product was only one small element of the product design. We carefully chose the airplane type that is allowing us to explore profitable markets that our competitors cannot serve efficiently with the larger airplanes that they use.

Another point discussed in this book is the innovation in the delivery of the services, both on the ground and in the air. We took a cross-functional integration within the airline to develop a "holistic" view of our customers, as discussed in this book. We also look at best business practices from other industries. We do not want to be too focused on conventional thinking in general and/or the conventional experience within the airline industry. The airline business is becoming a retail business and learning from this industry was and is very useful to Azul. We studied in detail companies that make efficient usage of information and technology to improve their operations.

Using business intelligence, as discussed in this book, we can develop novel ways to interface with our clients, travel agencies and the marketplace in general. We learned that this market is "unbalanced." More traffic is generated through businesses than from individuals, and a sizeable chunk of our business comes from travel agencies. As such, being able to communicate with every travel agent in contact with the customer is an important differentiation introduced by Azul. Unquestionably, it is also a huge challenge. The agent is in touch with the client and needs to sell the product differentials, not only the low fares associated with Azul. And to reach these clients, we need to pass through the travel agency. Another way to reach the customer is through social media, again, as discussed in the book. Azul has developed its presence in the main social media websites and has developed its own website, Viajamos.com.br. This presence helps to communicate with customers and save costs in advertising.

On the other hand, airline executives need to find a good balance between information, technology and intuition. A firm grasp of information and technology is mandatory today. However, sometimes, an executive needs to exploit his intuition and be bold enough to take risks, even if the numbers do not justify a decision. Take, for example, Azul's decision to fly between the cities of Campinas and Salvador. Before we started services, traffic data showed that only 35 passengers flew in that city-pair each day. Yet, Azul executives made a decision to fly nonstop with many seats per day and now there are 500 passengers per day. This proves that while information and technology are fine tools for decision making sometimes it is experience, judgment, and "gut feel."

Barueri, São Paulo, Brazil
June 2011

Foreword

Azran Osman-Rani
Chief Executive Officer
AirAsia X

As a (relative) outsider to the aviation industry, I am often amused at the near-obsession that many industry participants have with airline business models: Full Service, Low-Cost and even colourfully-named hybrids (from 'New World Carriers' to 'Five-Star Value-Carriers'!). Intense debates go on as to which business model or strategy is more successful than others, or whether there are 'sacred principles' of 'proven' low cost models that should never be challenged. Some are incredibly zealous about it – steadfastly declaring that their airline is superior because they strictly follow the established tenets of 'the low-cost bible' while others are doomed for deviating. Interesting that for all of the chest-thumping, these airlines are not even the lowest-cost operators!

I've long argued that business models are irrelevant. If they were, then all the low-cost carriers that adopted every single 'principle' of the Ryanair model should succeed. Many tried and failed. Similarly, if business models were relevant, then every full-service airline that emulates the Singapore Airlines model should succeed. They don't. To me, what matters are only two things: how well an airline executes (its much better to be a great executor of a mediocre strategy than a poor executor of a brilliant strategy) and implements its business, as well as the specific market context (supply and demand) where the airline chooses to operate (first mover advantage is almost universally insurmountable unless the first-movers royally screwed up in their execution).

For example, many derided us when we launched AirAsia X. Apparently, just because others had previously tried and failed to launch low-cost long-haul airlines, then the entire 'business model' is written off. Yet, in just three years and only a small fleet size, we've grown AirAsia X rapidly, doubling passenger traffic on routes we fly, achieving the world's lowest unit cost performance at 2.9 US cents per available-seat-km in 2010, and turning a profit. The difference, of course, is how we executed compared to the other airlines.

We also succeeded because of the specific market context here in Asia. Low-Cost Long-Haul is not seen as a priority for low-cost leaders in North America because they can cover virtually all of the continental United States with narrow-body aircraft. Similarly, the entire European market can be covered by narrow-body aircraft. For North American and European LCCs, going 'long-haul' is mainly about trans-Atlantic, a mature and saturated airline market. The situation is very different here in the more geographically spread-out Asia where one cannot fly from North Asian cities such as Beijing, Seoul and Tokyo to Singapore and Kuala Lumpur in Southeast Asia, let alone Australia, or from the Middle East and North India to Southeast Asia and Australia. These intra-Asian markets are also under-served today with high growth prospects driven by the demographic boom in Asia.

I therefore think its very timely for Nawal Taneja to address the specific areas that airlines should really be focused on, to compete and survive in today's highly volatile and challenging climate. Themes of customer-centricity and the use of new-generation technology to provide customers with choice and convenient services are the real basis of competitive advantage. We are big believers in investing to grow our global AirAsia brand, especially online and having a direct-to-consumer marketing and sales approach. Airlines that still choose to 'outsource' the customer relationship to third party intermediaries will do so at their own peril. Moreover, as airline leaders, I don't think we can delegate customer relations and just rely on market research reports to make executive decisions. The social media tools today make it very easy for us to stay in direct contact with our customers and have an immediate pulse on how our brand is alive – not in the advertisements we place but in the live conversations taking place among our customers.

Kuala Lumpur, Malaysia
June 2011

Foreword

Greg Schulze
Senior Vice President, Global Tour and Transport
Expedia Inc.

Most innovations are fostered during times of crisis. Emerging out of the depth of the recent recession, *technology* has once again become the leading driver of value creation. The evolution in technology—especially in mobile, social media, and search—is changing consumer interactions and preferences. The trend will continue, even accelerate in the next few years and will increasingly influence the travel industry. Nawal's latest book is an important and timely work that helps the aviation world better understand how technology will transform the customer experience, and once again, Nawal will help leaders form effective strategies.

About 15 years ago, Expedia began as a "start-up" within Microsoft. In the early days of the Internet, the success of Expedia and our competitors revolutionized the travel business and also gave credibility to the then nascent e-commerce industry. Not only did we increase transparency in the airline business, we stimulated demand... and of course informed and guided consumer behavior. While Web 1.0 was about information and transaction (e.g. Expedia.com) and Web 2.0 was about user generated content and virtual collaboration (e.g. Tripadvisor. com), Web 3.0 is largely about breaking down barriers between devices, omnipotent networks and specialized search. Web 3.0 perhaps holds the greatest promise to alter consumer interactions, and the change is already well on its way.

In the last quarter of 2010, smart phones outsold PCs for the first time in history. Globally, and especially in emerging economies, mobile usage and wireless technology have leapfrogged traditional infrastructure and societies have accepted mobile communication as acceptable business standard. For example, in the Philippines, ordinary citizens can pay their income tax using an SMS service on their mobile phone!

At Expedia, we consider mobile more than just another sales channel—we take it as a core part of our interaction with travelers. And mobile is no longer small. We have mobile sites in 80 countries in over 32 languages! We also have several apps across our many brands—hotels.com, Expedia.com, TravelTicker, Hotwire, Tripadvisor and Mobiata – many of which are amongst the top travel selections in the iTunes store. When Expedia launched its iPhone app for hotels in early 2011, there were 250,000 downloads within the first month, or an average of 10 downloads per minute!

Growth in mobile influences consumer purchase patterns and travel product features. For example - more than 75 percent of our mobile hotel bookings are for stay either the same night or the next night. Therefore, "dynamic geography based search" becomes a critical feature for the site/app, and "distance from the highway" becomes an important amenity for the hotel. Change is also coming fast in the airline world and this book will certainly help executives build a strategy that considers customer patterns changed by evolving technology.

While mobile is about how we access information, social media is about how we explore. Consumers, especially travelers, explore primarily based on opinions and experiences of others. Every minute, 21 reviews are added to TripAdvisor, the largest travel review site in the world. According to PhocusWright,[1] 3 in every 10 travelers share their travel reviews online, and 87 percent of consumers looking for a hotel today are influenced by reviews. On Expedia.com, we can see this influence in action - hotels with the good reviews of 4.0 or 5.0 generate more than double conversion than hotels with reviews of 1.0–3.0. What will this look like for airlines, as the travel experience becomes increasingly described by user-generated content? Again, Nawal's latest book gives informed perspectives that will help airlines better communicate their value propositions through emerging channels.

After gaining insights from social networks, the next step for consumers is usually to search for their specific travel options.

---

1   PhocusWright's "Social Media in Travel Planning", March 2011

Expedia has always been a leader in low-fare searches, and we continue to work hard to stay leading edge. The insights from Nawal's book are a strong encouragement for us all to efficiently provide relevant information to the traveler. The processing speed of computers is improving and machine-learning is here; searches can "intelligently" gather what you are looking for and instantaneously present the most relevant results. Further machine learning will enable even more relevant offers in the future based on what you clicked amongst the options presented.

Still, the single biggest driver for consumer search is the ability to comparison-shop. Some of my friends in the airline business presume that the evolution of search will provide them the opportunity to sell products in a customized manner for each individual, and thus allow them to differentiate their offering. This is only partially true — the core need of consumers to comparison-shop will not go unfulfilled. Evolution of search will not allow fragmentation of content; if anything, it will increase the value of comparison-shopping services in the eyes of consumers.

The age of constant connectivity and customers armed with highly relevant information has brought us new opportunities to engage travelers and to evolve our services. Our technology strategy and the ability to serve this new customer will determine the collective success of our industry. Readers of Nawal Taneja certainly have an advantage and should lead the creation of a better travel experience for the new consumer. We, at Expedia, are very excited about the prospects, and are looking forward to a brave new world for travel.

Seattle, Washington, USA
June 2011

Foreword

Francesco Violante
Chief Executive Officer
SITA

The best way to find out what passengers want from technology is to ask them. This is something the Passenger Self-Service Survey does every year at a range of the world's leading airport hubs spanning the globe from Atlanta to Mumbai, from Beijing to Sao Paolo, and visiting destinations in between such as Moscow's Domodedevo, Frankfurt, Johannesburg and Paris Charles de Gaulle. The survey provides a fascinating insight into how passengers are behaving now that so many of them have a piece of the air transport industry's IT infrastructure in the palm of their hands, in the shape of a smart mobile device.

As online booking reaches its full potential in many regions of the world, focus is shifting to mobility, the next horizon in the digital world when it comes to boosting airline and airport productivity and giving passengers ever more control over the steps of their journey, whether in the air or on the ground. The ownership of smart devices among the travelling public is growing rapidly. At 3 per cent, the percentage of passengers who had used mobile phone check-in on the day of the last Passenger Self-Service Survey remained the same as in the previous year, but the percentage who stated they had used mobile phone check-in in the past grew from 14 per cent to 23 per cent. Overall, 58 per cent of passengers are interested in using mobile check-in.

Beyond the convenience of being able to use a mobile phone to book your flight, check-in and to receive a bar-coded boarding pass and other flight-related information as required, what is it that passengers might expect an airport or an airline to provide through their smart devices? You don't have to look far for the answer. Next to their flight arriving on time, short queues are the second most important factor for a pleasant journey, according to the passenger survey. This is corroborated by the annual

Air Transport IT Summit for industry CEOs and CIOs where discussions are centred on themes such as coping with more passengers in the same terminal space, streamlining passenger processing and enhancing the overall passenger experience.

State-of-the-art developments are responding to the convergence of interests. A case in point is the recent innovative work at Copenhagen Airport. This involves the deployment of the world's first indoor augmented reality application which allows passengers to use the photo view of their smartphone to see information on gates, shops, restaurants and other services overlaid on their view of the airport in a fun and interactive way. This app uses triangulation and signal strength from Wi-Fi access points to determine the location of individual passenger's mobile phones (but does not identify the passenger). The airport can see the mobile devices represented in real time as dots on a 3D representation of the airport and can take pre-emptive action where bottlenecks look like being created.

As the author points out in Chapter 3, sophisticated real-time queue management at airports is high on the agenda for the industry. When combined with augmented reality, this gives rise to the possibility of a very close relationship between the passenger and the airport operator to their mutual benefit, with the more rapid deployment of resources to tackle lengthy queues at check-in, security, duty-free shops or boarding. The air transport industry needs innovative approaches like this to solve the issues it faces, while also working in close cooperation with IATA on critical industry-wide programmes such as *Simplifying the Business* and more lately its *Passenger Experience* initiatives to deliver efficiencies for all stakeholders across the air transport eco-system.

It is also vital that we work collaboratively at the cutting edge of digital innovation for the benefit of the entire air transport community. This has been the role of SITA Lab since its establishment three years ago. The industry is seeing promising innovations that point the way forward for all of us, such as the launches by Malaysia Airlines of the world's first kiosks to sell airline tickets using the Apple iPad, as well as a Facebook application that allows users to book and check-in for flights with full integration into their social graph. Before that,

Malaysia Airlines' MHDeals also became the first augmented reality application for an airline, allowing users to pick up the airline's best deals from nearby airports. A new payments service for the airline industry will support mobile and other booking applications. The author discusses game-changing innovations such as these in Chapter Two.

In the next two years, the vast majority of airlines plan to invest in IP broadband connectivity both to and from the aircraft and many have already done so, as evidenced by the growing number of OnAir customer airlines anxious to tap into the inflight market for digital and mobile phone communications, particularly in response to demand from business passengers.

The aircraft is becoming another "node on the network", a flying data centre linked through wireless broadband to an airline's ground-based network. High-speed upload capabilities will save time and introduce operational efficiencies in areas such as software and in-flight entertainment loading. The huge volumes of data generated by e-enabled aircraft will result in a plethora of advantages to passengers (and of course airlines and crew), not least of which will be improved aircraft turnaround and, for the environmentally-aware, the provision of the CO_2 data necessary to measure industry emissions.

These 'connected aircraft' are an important area of focus, along with digital travellers, intelligent airports, mobility, virtualization and cloud computing – as evidenced by our New Frontiers papers which consider innovations and technologies that will impact profoundly on the air transport industry over the next three to five years. They have flagged how quickly mobile devices, Web 2.0, Near Field Communications, RFID and biometrics are becoming mainstream as the industry strives to cope with huge growth in passenger volumes. In embracing transforming technologies such as these, we must be guided by the industry's need for reduced costs, greater competitiveness, improved profitability and operational performance. We must also remain focused on exploiting all engagement models with passengers, as well as exploring new revenue channels while enhancing the passenger experience – making air travel easier, safer and hassle-free.

One of the biggest game-changers for the air transport industry will undoubtedly be cloud computing. The provision of a shared Cloud infrastructure will be one of the great enablers for passengers in the digital age as they seek seamless connectivity and access to airline and airport applications with zero downtime. This technology will enable airline and airport systems' integration and a move to software-as-a-service applications that can be used from any connected device, anywhere in an airport or across different airports. Better integration of airport systems will mean that information flow to the passenger on flight or baggage delays, for example, can happen in real time.

Above all, passengers want a trouble-free journey. Smart mobility will provide airports with direct access to passengers for information on their distance from the gate, flight announcements, retail bargains, and delayed baggage. 70 per cent of airports plan to provide mobile services by 2013 when almost 50 per cent of travelers will be carrying smart phones. A robust airport communications infrastructure based on 4G and Wi-Fi technologies will provide seamless, permanent connectivity.

We are standing on the threshold of a digital world where airports and airlines will be able to leverage a passenger's location because of ever-present personal-area networking and sensor technologies, such as Bluetooth and RFID, wireless LAN, and 4G products. On a practical level, baggage is one example of how this might work to everyone's benefit. Baggage is always a major concern for passengers if either they are late to the airport or the flight is delayed. In the near future, airlines will be able to make more real-time, informed decisions on whether to load or off-load bags based on the passenger's location. A passenger who is not at the gate will be contacted directly via their mobile device and instructed to make their way to the gate for boarding. However, if advanced analytics predict that a passenger will not be able to board on time (based on their location), the passenger and bags will be automatically rebooked allowing the flight to depart on-time. Double-win scenarios such as this for both passengers and airlines will inevitably result from passengers going digital and mobile. Otherwise it would be extremely difficult for the air transport community to cope with an expected increase of 800 million passengers over the next four years.

Geneva, Switzerland
May 2011

Acknowledgements

I would like to express my appreciation for all those who contributed in different ways, especially, Angela Taneja, an experienced analyst of best global business practices, Peeter Kivestu (Director, Industry Marketing and Solutions for Travel), Dr. Dietmar Kirchner (formerly with Lufthansa and now a Senior Aviation Consultant), and Rob Solomon (Senior Vice President and Chief Marketing Officer at the Outrigger Enterprises) for discussions on challenges and opportunities facing the global airline industry and related businesses.

The second group of individuals that I would like to recognize include, at: Accenture—Guido Haarman and Anja Wickert; Air New Zealand—Stuart Hilton and Jodi Williams; AirTrav Inc.—Robert Kokonis; Amadeus—Julia Sattel; American Airlines (formerly with)—Scott Nason; Boeing Commercial Airplanes—Fariba Alamdari; Bombardier Aerospace—Philippe Poutissou; Centre of Asia Pacific Aviation—Peter Harbison; Continental Airlines (now part of United Continental Holdings)—Chris Amenechi, Scott O'Leary, Alex Savic, and John Slater; Delta Air Lines—Chul Lee; Ethiopian Airlines—Nigusu Worku; Expedia—Greg Schulze; Forrester Research—Henry Harteveldt; Google—Rob Torres; Hawaiian Airlines—Richard Peterson; IATA—Martin Braun; IBM—Eric Conrad; IRC-Aerospace—Tomoo Nakayama; Lufthansa German Airlines—Christoph Klingenberg; MMG Worldwide—Clayton Reid; SITA—Juergen Koelle, Ian Ryder, and Terence Tucker; Southwest Airlines—John Jamotta and Pete McGlade; Teradata—David Schrader; TripIt—Gregg Brockway; and Ypartnership—Peter Yesawich.

Third, there are a number of authors whose work and ideas have been referenced numerous times in this book. They include Ken Auletta, Gillian Jenner, Mary Kirby, Scott Klososky, Rick

Mathieson, Ron Reed, Jeanne Ross, Nick Smith, Peter Weill, Robert Wollan, and Catherine Zhou.

Fourth, there are a number of other people who provided significant help: at the Ohio State University— Bradley Hock, Alex Holmes, and Jim Oppermann; and at the Ashgate Publishing Company (Guy Loft—Commissioning Editor, Kevin Selmes— Production Editor, Peter Stafford—Proofreader, and Luigi Fort— Senior Marketing Executive).

Fifth, I would like to acknowledge the help of Brian Wilson, Marne Miller and Jason Kimerling at Wilson Advertising for the design of the jacket for this book as well as the images contained in the Appendix in this book.

Finally, I would also like to thank my family for its support and patience.

Introduction

To its credit, despite the overwhelming institutional constraints and complexity of its operations, the global airline industry has achieved remarkable success in providing a safe and reliable air transportation system. And, despite the fact that the industry has not made an adequate return on its invested capital (over a long period of time), it has managed to optimize its operations to adapt to the changing environment—various phases of the liberalization policies around the globe and the emergence of low cost airlines around the world, for example. The conventional network carriers optimized their operations relating to hubs, fleet, networks, and alliances. For their part, the low cost carriers also began to optimize their operations by increasing the number of conventional airports served, increasing the number of aircraft types in the fleet, and are beginning to interline with other low cost carriers as well as the conventional network carriers. While these changes to the airline business models worked reasonably well in the first decade of this century, the traditional optimization framework (networks, fleet, schedules, alliances, and so on) is not likely to be sufficient to deal with the increasingly powerful forces, in motion and in the early stage of formation.

To begin with, customer expectations are rising at an exponential rate, leading to the need for airlines to shift their focus from operations and product centricity to customer centricity. The reason for the change in emphasis is not based on the assumption that product and operations centricity are not important. Rather, airlines have already made significant progress in these areas. Now, there is a greater need to shift focus to customer centricity. What does that mean? Customer centricity means different things to different passengers. For some, it may mean implementing processes that deliver an enhanced customer experience. For other passengers it may mean implementing processes that give more control back to passengers. For example, it may mean providing options that suit their situations and needs, especially in times of disruption/irregular services. Customer centricity also varies by generation, for example, with respect to tech savviness.

In addition to the changing needs of the changing customer base, new forces also include the emergence of well-financed and hypercompetitive airlines based in the Gulf region of the Middle East, the three fast-growing airlines in China, and now the advancements in technology, exemplified by the price-service comparison websites and user-centric rating websites.

- Consider the rate at which airlines with different business models are entering the marketplace. Ten to fifteen years ago, there were the conventional network carriers such as United, British Airways, and Qantas. They competed among themselves as well as with standard low cost carriers such as United with Southwest and JetBlue in the US, British Airways with Ryanair and easyJet in Europe, and Qantas with Virgin Blue in Australia. The spectrum of low cost carriers with different business models now includes the Las Vegas-based Allegiant, the Barcelona-based Vueling, the Sharjah-based Air Arabia, the Kuala Lumpur-based Air Asia (and its subsidiary Air Asia X), the Singapore-based Tiger Airways and the Sydney-based V Australia (a division of Virgin Blue). Airlines entering the marketplace with new business models cannot be ignored by the conventional network carriers. Let us not forget what happened to global icons that relied on incremental changes when new players entered their business, whether the icons were airlines (such as Pan Am) or non-airlines (such as AT&T).
- As already mentioned, consider the rate at which consumer expectations are changing, based partly on the needs and values of new generations (shaped, in turn, by technologies such as mobile devices and social networks), and partly on the experience provided by non-airline businesses. Information and consumer technologies are changing passenger behavior and passenger expectations as well as expectations of employees. Passengers now have richer and

timelier information, enabled by digital, mobile, and peer-to-peer interaction through social media. Passengers are now empowered by tools and information, price-service comparison websites, and also user-centric rating websites. Keep in mind the insights from businesses that transformed their business models to meet changing consumer expectations, may they be airlines (such as Virgin America and Allegiant) or non-airlines (such as Apple).

- Consider the potential impact of the entry of new technology players in the airline business who are ready to deploy game-changing technology (such as "smart" search and shopping engines, "smart" mobile devices, and "smart cards") to finally be able to operationalize such concepts as mass customization with their companies making money by becoming providers of solutions and value integrators. While suppliers (for example, airlines) and distributors (for example, GDSs) debate challenges and opportunities relating to direct and indirect channels, new technology companies could enter the marketplace and change meaningfully the distribution of airline services. These companies will develop truly customer-centric business models founded in information and enabled by technology. For example, before an offer is made, they will first find out sufficient and relevant information on the person seeking information on the services desired. They will then provide an information-based customized offer rather than simply the list of flights available and the listed fares. Again, consider the impact of new players on incumbents, may they be airlines (such as Emirates on British Airways and Lufthansa through the use of integrated strategies and value-adding price-service offers) or non-airlines (such as Sprint on AT&T and more recently Google and Android on Nokia and Symbian).

These powerful changes are necessitating major changes in airline business models. Airlines are clearly aware of the major trends and their implications and have begun to change their business models. Consider, for example, the rate at which the business models of the two major categories of airlines (low cost and full service) are converging. Low cost airlines have begun to serve large conventional airports, interline within their own sector as well as with conventional network airlines, start flying in intercontinental markets, distribute their products through traditional GDSs, and offer two-class service in long-haul markets (for example, Air Asia X and V Australia). Consider also the speed at which traditional network carriers have been growing their ancillary revenue following the lead of low cost airlines and developing ways to compete more effectively with low cost carriers (Qantas through Jetstar, Singapore through Tiger, and Iberia through Vueling). However, the forthcoming changes are so dramatic that they will require greater innovation in the business models than optimization along the traditional dimensions. Information would most likely be the foundation of customer-centric innovation to meet the needs of different generations of travelers with different socio-economic backgrounds based in different regions of the world. Such information will play an important role in identifying what some business analysts are calling, "the edge of the company and the edge of the market."[1] And technology could enable some conventional network carriers to reinvent their business models to compete with some low cost carriers and to cooperate with others.

Customer-centric innovation based on information will be enabled by the new-generation technology. Ironically, it is also the new-generation technology that is raising the consumer expectation bar, for example, the demand for control, personalization, and instant gratification. The question then is: Can airlines make use of the new-generation information and enabling technology to meet the step-changing expectations of a very broad customer base? This book reviews (a decade after the first book) the relevant evolving information and enabling technologies to identify and implement flawlessly new business strategies. It is the deployment of new information that will help management identify new points of integration where value can be added.

- The first chapter contends that, despite the inherent constraints under which the global airline industry functions, it is possible for airline CEOs and CIOs to transform their business models from those that have focused on economic and efficient air travel from airport-to-airport to those that focus on solving different problems faced by different passenger segments, through cost-effective mass customization and value integration. While these concepts, such as personalization, are not new, now their implementation has become feasible due to the availability of new-generation information and enabling technologies. For example, relevant and timely information, coupled with new-generation technology, can now not only enable an airline to develop, sell, and charge an appropriate price for a personalized and solution-based service, but also to deliver it in a personalized manner for premium-fare passengers. For the larger segment whose behavior is shaped mostly by price an airline can develop services that match fares and services offered by competitors.

• The second chapter provides a glimpse of the direction of new technologies enabling the identification of information and the implementation of business strategies for airlines to provide solutions for different segments of air travelers. Examples of the new-generation technologies include incredibly advanced shopping engines (with the capability to synthesize browsing and buying behavior), sophisticated mobile devices (smartphones such as iPhones and BlackBerries as well as tablet PCs) applications and advances in the development of virtual assistants, to provide personalized and solution-focused services. There is also the new wave of social technology (the latest networks, sites, and tools) and context-aware and location-based mobile applications to develop game-changing marketing initiatives. Examples include advertising banners on a website that change to adapt to the browsing/buying behavior of the individual, and personalized MMS (multimedia messaging service) that include multimedia material (audio, video, and images). Next, evolving advancements in the capacity, speed, and cost of computers coupled with the capability to develop a single large database with data mining and data analytical functionalities can provide airlines the means to finally achieve *integrated planning* at much higher speeds, while producing results that are much more robust. The chapter also provides three mini case studies of best global business practices: Groupon, Sephora, and Unilever.

- The third chapter provides some examples of information-based and technology-enabled initiatives that some airlines have already implemented and are implementing as of the beginning of the second decade of this century. Examples of such initiatives include mobile check-in (including the transmission of bar-coded boarding passes), website virtual assistants, location-based smartphone applications, and limited endeavors toward providing door-to-door service rather than just gate-to-gate service. This chapter also discusses how some airlines are excelling in the area of social media to further engage passengers. However, airlines can go still further to satisfy even a few customers who have gone on to establish their own websites (tools created by travelers for fellow travelers in areas where service has otherwise fallen short). There are some strategic insights presented for airlines. First, passengers are going online for their products and services at an incredible rate. Second, within the online framework, the mobile channel is growing exponentially. Third, and perhaps, more important, airlines should not only focus on these channels from the viewpoint of shopping, but also from the viewpoint of brand communications (and some have already begun to do so). The chapter provides three mini case studies of best global business practices: Interactive Fitness Holdings, OpenTable, and Zipcar.

- The fourth chapter touches upon some areas that are ripe for further passenger-centric innovation. This discussion on the possible information and technology-led innovation through meaningful passenger engagement is presented in four parts: (1) passenger segmentation, (2) passenger relationship management, (3) passenger loyalty, and (4) flawless passenger experience. The important part in the segmentation process is the need to understand what motivates each customer's behavior. As for customer relationship management the fundamental point is to recognize what is important to a passenger—information that can change in different situations for the same passenger. The flawless execution process requires a solid integration not only among different functions of one airline, but also across alliance partners to provide passengers with an authentic seamless experience. The fundamental points relating to passenger loyalty are getting the basics right, providing passengers with meaningful solutions, and developing and maintaining credibility. One key point is the recognition that these areas of passenger-centricity are not independent. The integration of these parts could make an airline a better value integrator and a solution provider. The chapter provides three mini case studies of best global business practices: BMW, citizenM, and megabus.com.

- If game-changing information and enabling technology are available and if, for the most part, some airline managements have begun to recognize the value of information and technology in helping them find solutions to the problems of their strategically- and intelligently-segmented customers as well as their own problems internally, then what are the hurdles in the implementation of such information and technology? Chapter 5 discusses numerous major hurdles: (1) inadequate business intelligence, (2) lack of actionable metrics, (3) overlooked integration opportunities, (4) lack of technology strategy and investment plan, (5) inadequate knowledge of social technology/media capability, and (6) non-alignment of business and enabling technologies. The overarching challenge lies in the fact that the whole planning perspective appears to be unfocused and fragmented.

- Chapter 6 provides some opportunities for overcoming the hurdles in the implementation of the new-generation information and technology. The basic assumption is that these hurdles currently lead to an unfocused technology strategy and the investment in technology. Therefore, airlines need to sharpen their focus by implementing strategies to overcome these hurdles. For instance, instead of the amorphous perspective of business and technology strategies that has existed in the past, an airline needs an approach which is as sharp as a diamond, focused at the point of competitive impact, strong enough to overcome competitive resistance, and shine in terms of customer satisfaction. Information strategy is at the top of the diamond, while technology strategy is at the bottom. Business strategy (focused on customer centricity and complexity management) is on the left-hand side, while new points of integration (leading to new ways of creating value and profit) is on the right-hand side of the diamond. The chapter also provides some practical examples of these strategies from outside the airline industry.

- As mentioned throughout this book, while new-generation information and enabling technologies can be the centerpieces of radically different business model initiatives, they need to be evaluated and deployed in a holistic framework to meet the changing needs of changing customers. The alignment among information, business, and enabling technology strategies calls for a five-step action plan by management, discussed in Chapter 7, the concluding chapter. The common ingredient among the five-step action plan relates to the responsibilities of the CEO, the responsibilities and accountabilities of the CIO, and the organic interaction between the two executives.

- Finally, the Appendix contains an insightful case study created by Teradata suggesting a complete overhaul of an airline's rebooking application to improve its customer satisfaction ratings.

Even the new-generation information and enabling technologies cannot solve some structural problems faced by the airline industry (such as the elimination of antiquated government regulations and intervention as well as problems caused by inclement weather and insufficient aviation infrastructure). However, they can provide cost-effective solutions to strategically-segmented passengers' problems (thereby improving immeasurably passenger service and passenger experience). They can also enable employees to have access to information and tools to meet varying passenger expectations at all touch points. And, they have the potential for airlines to make reasonable margins through business cycles. What are needed are strong-willed and visionary CEOs and CIOs to develop innovative and customer-centric information and business strategies that are executed flawlessly.

Notes

1   Paul Nunes and Tim Breene, "Reinvent Your Business Before It's Too Late," *Harvard Business Review*, January-February 2011, pp. 80–7.

Chapter 1
Creating Information-Driven and Technology-Enabled Solutions for Airlines and Passengers

There is a game-changing shift taking place in the business sector worldwide in which firms are transforming the way they conduct their operations and the way they respond to their customers' needs, at any time and at any place and in, as close to as possible, personalized ways. This game-changing shift is being driven by new-generation information incorporated in the business model and enabled technology, described in the next chapter (such as personalized search engines and mobile technology, including applications). Technology can now enable the collection and analyses of relevant, timely, and personalized information to become a key element of the business strategy, enabling a business to become truly customer centric and to provide solutions and aggregate value for the customer at a personalized level. At the same time, the new-generation technology is enabling firms to reduce their costs, enhance their revenues, reduce their business risks, and become much more flexible and agile by managing complexity. Moreover, new-generation technology is enabling firms to leverage their old and new sources of data to improve both their internal productivity and their strategic external collaboration to provide greater value for customers.

Mass Customization

Concepts such as customer centricity through mass customization and businesses becoming providers of solutions as well as value integrators have been around for many years. Take the concept of mass customization. It was discussed in books dating back to the 1990s.[1] The idea was to explore if products and services could be customized and personalized for individual customers but produced at low costs, leading to low prices. Some businesses viewed the concept to be contradictory based on the supposition that mass production can lead to low prices, but only at the expense of the products and services being uniform. However, some businesses proved this assertion to be invalid by implementing the concept successfully, for example, the computer company, Dell, and the insurance company, USAA. Consequently, while neither the concept nor its implementation (despite on a limited basis) are new, it is the need for its mass market penetration that has become more essential, including within the airline industry, for at least five reasons:

- Passengers' lives are becoming complex, time-constrained, and more diversified. Therefore, passengers are looking for easy solutions for their sometimes complex travel needs, not transactions (even if they are efficient), or even the lowest price.
- Competition is increasing among existing airlines as well as from new players, such as technology companies that are about to take distribution to new levels through intelligent market segmentation and personalized offers.
- The high-margin segment of the marketplace (premium travelers) is growing at a much lower rate than the lower-margin (leisure travelers) and it is becoming more discriminating. The lower-margin passengers, on the other hand, while benefiting from inflation-adjusted declining fares, are experiencing various levels of frustration relating to the services provided and lack of sufficient transparency. Middle classes are increasing in many emerging markets and their needs are different than the needs of the middle classes in developed markets. Populations are becoming older and living longer.

- Business is shifting from the West to the East with high growth rates expected in Asia Pacific and lower in North America and Europe.
- The airline business is becoming even more complex with the convergence of business models among the two main sectors within the airline industry (low cost airlines and full service airlines) and the increasing uncertainties and volatilities relating to the global economies and all sorts of other challenges (like volcanic ash, oil price fluctuations, or environmental restraints).

Ironically, just as the need for mass customization is becoming more imperative, the availability of a new-generation of information (for example, Web traffic and interactions through social media) and technologies (particularly, online, mobile, and social) now make information-based mass customization feasible, providing not only variety with a lower production cost, but also a higher quality. The higher quality feature can come from the delivery of services that meet individual needs and provide a better buying experience (enabled, for example, by the comparative shopping feature). At the same time, the successful implementation of cost-effective mass customization can enable airlines (now offering commoditized products/services) to be able to identify diverse segments and provide each one with a differentiated value proposition, thereby moving the focus away from buying on price.

The main thought to keep in mind is that new-generation information and enabling technology are now providing both passengers and airlines with an unprecedented number of choices. For passengers, new-generation information provides service options relating to airlines, airports, fares, alliances, and so forth. For airlines, there are options to use technology to cut costs, introduce new functionality, reduce time to market, or improve customer experience. The whole idea behind the airline business model transformation is for people in the airline organization at different levels to be able to obtain relevant and timely information and convert it into intelligent solutions for passenger-related or airline-related problems in a quest to become more customer centric.

New business model opportunities can be explored from the intersection between three components: (a) global megatrends and airline industry foresights, (b) passenger insights and their unarticulated needs, and (c) new-generation consumer technologies and capabilities, as shown in Figure 1.1. First, there is competition between low cost carriers and the large network carriers. The latter group is coming to the conclusion that while it has reduced its operating costs, the cost levels are still too high to provide competitive services in short- and medium-haul markets. Within the low cost sector it is also becoming evident that there are limits to the future growth of airlines if they were to continue with their current business strategies. This point is addressed later in this chapter. Second, competition will increase at an unprecedented rate for the three major European carriers as the three major Gulf-based carriers in the Middle East increase their capacity. Emirates Airline has already become one of the top 10 carriers in terms of its size, operating to 65 countries with a fleet of 150 wide-body aircraft. More than 200 aircraft are still on order. Third, competition will also increase for the old world global airlines from the four major airline groups based in China — Air China, China Southern, China Eastern, and Hainan. According to the research conducted by the Centre for Asia Pacific Aviation, Air China is now the world's top valued airline in the eyes of investors.[2]

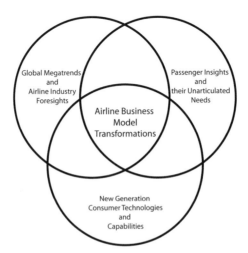

Figure 1.1   Potential Framework for Airline Business Models

In terms of passenger insights, airlines can now include input from passengers not only in the development of new service offerings, but also in creating solutions to their current problems occurring at most touch points in the travel chain and improving the passenger travel experience. Most importantly, airlines can now implement ways of capturing both articulated and unarticulated needs by monitoring, for example, Web shopping behavior. These needs vary significantly by passenger and by situation for the same passenger. Even relating to a given need, such as greater transparency, there is much variation. One passenger wants greater transparency relating to the fare while another wants transparency relating to information that is sometimes inaccurate, untimely, or irrelevant from the point of view of the ability to take an action. Some airlines are beginning to offer passengers the ability to customize their travel experience. EasyJet in Europe was one of the early adapters of this concept. Figure 3.3 in Chapter 3 illustrates United Airlines' initiatives in this area. Clearly, there are limitations relating to the size of the market segment desiring customization, the limitations of infrastructure to produce customized products, and the costs of providing customized products. Admittedly, infrastructure is also required for information and technology. And, this infrastructure is not trivial, as it requires planning, growth, and investment, just like any other kind of infrastructure. However, unlike traditional infrastructure, that becomes less useful over time, information and technology infrastructure, especially when it is centered around data collection, becomes more powerful and more broadly applicable over time.

The airline industry foresights component relates to a comprehensive understanding of emerging megatrends and monitoring their continuous shifts to develop new competitive strategies. These foresights relate to various forms of consolidation and various forms of cooperation and competition among carriers within alliances and between different sectors of airlines. New-generation consumer technologies and capabilities (such as personalized search engines and digital travel concierges) relate to the acquisition of appropriate information for the first two components and for airlines to act accordingly with proactive, dynamic, and agile passenger-centric and competitor-centric strategies.

Information-based mass customization, assuming it is one way to transform airline business models, requires not just the acquisition of new-generation information and technology, but it will also require a clear articulation of vision, a fitting organization structure, an enabling corporate culture, adaptive supportive systems, and expediting processes. Vision relates to how an information-savvy airline might see a creative mass customization technology platform to achieve profitable growth that is sustainable. Organization structure refers to not just where the technology group sits but also how it facilitates communications, coordination, and collaboration among closely related groups such as those shown in Figure 1.2. Systems refer not just to technology but also to how employees are selected, trained, empowered (provided with business intelligence tools), kept motivated, and rewarded. Processes relate to how customer profiles are built using behavioral data, how they are kept updated, and how they are used to provide solutions to passengers.

In the final analysis, information-based mass customization would depend on the availability of relevant and timely information that must be available to passenger-facing employees at each touch point (at check-in or in flight) on a consistent, personalized, and actionable basis. This information must also be available through a wide array of devices and channels (websites, kiosks, call centers, mobile devices, and so on) and at different locations, including in the aircraft itself, while in flight, so as to be able to engage with an individual passenger, and provide meaningful solutions to problems as they occur, or even better, before they occur.

In addition to the passenger-facing staff, executives also need to have access to technology and information to test new mass customization ideas themselves, and quickly, rather than depend entirely on the centralized technology department. Given the importance of this capability, it is the responsibility of the technology department not just to acquire and implement new-generation technologies but also to support the development and assessment of new, including ambiguous, concepts on an ongoing basis. Information-based mass customization clearly requires on an ongoing basis business intelligence (including the functionality to integrate and synthesize unstructured data) on internal operations as well as external parties such as passengers, competitors, and vendors as well as authorities and regulators (as shown in Figure 1.2). Therefore, planning processes need to change dramatically in all departments to achieve mass customization, enabled by information and technology, to take a deep dive into the purchase behavior of passengers and their expectations for solutions when problems arise. While such a deep dive can easily be achieved by mining information through the use of such technologies as Web crawling and text mining, the achievement of mass customization requires also a change in corporate culture to break down the internal and external silo systems, typified in Figure 1.2.

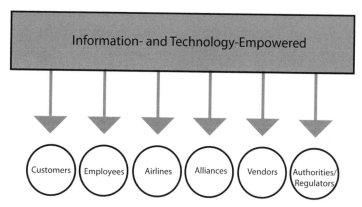

Figure 1.2   Information- and Technology-Empowered Holistic Framework

Solution Orientation

Just as with mass customization, the story is similar in the case of companies becoming solution providers. One famous case study relates to the leadership of Louis Gerstner Jr. when he was able to turn IBM around, in part, by repositioning a company that sold computer hardware, software, and information and technology services to a company that provided solutions to its customers' problems. While the concept itself is fairly straightforward, its implementation is complex as it involves changing the culture of the corporation. Going back to the case of IBM, it was a strong-willed outsider who was able to change the culture. For example, insiders were not willing to accept defeat with the marketing of some products, such as, the OS/2 software and the IBM desktop computers. At the same time, the three major internal divisions remained focused more on internal competition than on the needs of external customers. It was Lou Gerstner's success in changing the culture of the corporation that enabled the successful implementation of the concept of a solution provider.[3]

Can such a cultural change be implemented in many large "old world" airlines to build a stronger relationship with their passengers to become proactive even in creating solutions, let alone creating customized solutions that are innovative? The time to make the step-changing culture shift is now given that passengers are demanding new service value propositions, and information and technology are now becoming available for airlines to interact with passengers in real time to learn about their real problems and provide meaningful solutions, even if it means recommending and selling the services offered by competitors. Should an airline sell a seat of its competitor on its website if it solved the critical need of a passenger (even one of its top-tier passengers, let alone a new passenger)? Ironically, there are some airlines that are reluctant to even sell seats on their alliance partners if it is possible to keep the passengers flying on their own aircraft even if their own flights are less convenient. This predicament is an example of some airlines' lack of passenger-solution perspective.

As with the case of mass customization and provision of solutions, the concept of companies becoming value integrators is not new either. In fact, within the aviation industry itself, Pan American World Airways was the first airline to exploit this concept. The legendary airline entrepreneur and founder of Pan American, Juan Trippe, made transatlantic and transpacific travel possible when there were no viable airplanes, no qualified crews or crew bases, no navigational aids, no en-route landing fields, not even bilateral agreements between governments. Even if some of these products, facilities, services, and regulations had existed, true value could not have been achieved by simply integrating them. He was a true value integrator in that he created the missing elements, coupled them with the existing elements, and then operationalized the entire combination at the point of integration, *and that is how he made money and made Pan Am a global brand.* Decades later, FedEx succeeded in creating a highly demanded value proposition (door-to-door delivery of small packages) by starting with the integration of the services provided by trucking companies with those provided by airlines. However, real value was provided not simply by combining the services of two different transportation businesses, or even by utilizing the hub-and-spoke system (that had previously been used by airlines such as Delta and Eastern in Atlanta), but by integrating the combination of the three with a fourth component, the creation and deployment of information and technology, first to operationalize the door-to-door concept, and later to provide even more value by enabling shippers to track their packages.

Accordingly, application of the aforementioned concepts (while not new to businesses, including the airline sector) can now be taken to new heights through the use of new-generation information and enabling technologies. Admittedly, in the case of the airline industry, their implementation has been somewhat difficult because of the need to accommodate on one airplane a very broad spectrum of passengers (with an equally broad spectrum of needs and sensitivities to price) and a wide array of operational constraints, with many outside the control of the airline industry. Despite such difficulties, some airlines have already proven (to a limited extent, given the operational constraints) that they can become value integrators and solution providers.

Let us first consider some historical examples of airlines being solution providers for passengers:

- Passengers wanted to travel from their points of origin to their points of destination on multiple airlines (flying in multiple segments) using one ticket issued by one organization (an airline or an agent) and paid for through one transaction (say a credit card) and in one currency. Airlines developed interline agreements, bank settlement payments, and a single ticket for multi-segment travel.

- Passengers living in smaller cities wanted higher frequency service in a broad spectrum of markets and at reasonable fares. Airlines developed hub-and-spoke systems that connected thinner markets among themselves as well as with higher density markets and with a broad spectrum of fare levels and structures. Subsequently, global hubs were developed to take the solution to the passengers' problem to new heights, providing, for example, multiple routings for a passenger from Albany, New York to Zagreb, Croatia.
- Passengers wanted a much broader spectrum of fares. Airlines developed revenue management systems that not only solved this problem but also found the systems extremely valuable for themselves to increase individual flight profitability.
- Passengers were getting restless on ultra long-haul flights. Airlines developed in-flight entertainment systems (including individual systems at individual seats) and, more recently, the in-flight Internet-based services to keep passengers entertained and productive.
- Passengers wanted seamless travel and airlines wanted to access different markets which were restricted to them due to the existence of constraining bilateral agreements. Airlines developed strategic alliances. A passenger could buy a ticket on one alliance partner and receive, more or less, seamless travel—for example, boarding passes for connecting flights.
- Passengers wanted to rest on long-haul flights, especially business travelers who needed to arrive at their destinations to start work almost immediately upon arrival. Airlines developed lie-flat beds. Emirates now even provides on-board shower facilities on some aircraft on some routes.
- Passengers wanted more control of the check-in process and airlines wanted to reduce their check-in costs. Airlines facilitated the check-in process with self-service machines at airports and boarding passes through remote printers as well as via mobile phones using bar-coded systems. The best example is the introduction of the Web check-in capability.

Take the case of airlines being solution providers for their internal challenges.

- Airlines wanted to show their contents worldwide and access passengers worldwide. The solution was provided by the deployment of GDSs.

- Airlines developed fairly sophisticated systems operations centers that restored airline operations after major disruptions caused by, for example, the closure of airports due to severe weather. Unfortunately, the time taken was too long to put the system back together and, given the complexity of the problem, attention was focused initially on two legs of the stool, aircraft and crews, with less attention devoted to the third leg, passengers. Information and technology are now enabling airlines to reduce the crank-up time while integrating the needs of all three components— aircraft, crew, and passengers.

- Although airlines developed analytically-based planning systems that not only were capable of explaining what happened, but also what was happening, they faced three problems. First, the analyses were performed on a piecemeal basis (due to the continuous existence of the silo system and the lack of a common IT platform), producing sub-optimal results. Second, cycle times were high and reaction to problems slow. Third, systems were not able to predict a robust outcome if the airline continued to follow the adopted strategy. Now, a combination of information and technological advancements in numerous areas are being considered to produce results at a more rapid pace and results that are optimal on a system-wide basis.

- Airlines wanted to improve their flight-deck and line-maintenance productivity. This benefit has already begun to be realized through the implementation of electronic flight bags (EFBs) and electronic technical logs (ETLs). There are various levels of EFBs and their benefits can range from helping flight crews in filling forms and in reporting, to receiving data from an airplane's avionics systems to make performance calculations and, with advanced systems, to receiving information on and reporting technical problems. The ETLs relate to paper flight technical logs and reports that need to be sent to maintenance control centers and line maintenance departments. This information can be transmitted while the airplane is in flight.[4] Consequently, the increasingly e-enabled aircraft are helping airlines to improve their operations relating to dispatch, flight, and maintenance.[5]

However, despite these exemplary historic initiatives to be value providers and solution providers, in the fundamentally and structurally changing landscape of the global airline industry, the increasing expectations of consumers (based on the high level of progressive experience provided by products/services from companies such as Apple and Google), the increasing number of available options, and the use of a new wave of social technology, it has become much more critical for airlines to take the implementation of the aforementioned concepts to new heights. For example, while some airlines did add value at the points of integration (development of interline agreements, hub-and-spoke systems, and alliances), and did make money, they must now look for the new points of integration, now that all airlines have followed the same strategies. Airlines now need to find new points of integration to solve their passengers' problems creatively and to make money, with strategies being at least enabled, if not driven, by the new information and technologies. For example:

- Some passengers want smart search and shopping engines that are truly personalized and customized as well as information that takes into account past buying behavior and the availability of websites that, at least, provide transparency, let alone competitors' price-service options. For their part, airlines need systems that offer predictive outcomes for products and services being offered, as well as systems that provide explanations to their potential customers of why their services are better. For example, one airline may show a higher fare but may offer, not just a lie-flat seat, but a horizontal seat, a power source on the seat to connect a laptop, a truly efficient airport transfer lounge with knowledgeable staff and facilities for passengers making connections at international airports, "live" agents on the website that answer specific questions (agents who either have answers or can get answers, accurate, "around the clock," and in any area). Currently, most airlines are not at this level of sophistication. Unfortunately, a few are not even able to provide answers to even basic questions. An international agent within the call center of a global airline should be able to answer the question as to whether a passenger arriving in Terminal 2 on an international flight at Frankfurt Airport in Germany could transfer to Terminal 1 while remaining in the secure area. The passenger should not have to access the website of the airport or other sources (such as colleagues who have gone through such a process) to obtain an answer to her question. A passenger checking in at an airport in Africa to connect with an international airline in Europe should be able to get a boarding pass for the connecting flight, especially when the connecting airline is a strategic alliance partner, even though the connecting flight in question may not be a code-share flight. Keep in mind that *seamless travel* continues to be very high on the priority list of passengers.

- Some passengers want 24/7 travel concierges that follow passengers from start to finish and offer solutions for all kinds for en-route problems (across the whole travel chain, across all airlines and not just within a given alliance, and across all travel partners such as hotels and car rental companies). Passengers also expect airlines to develop comprehensive mobile strategies in which a broad spectrum of mobile devices can be connected to systems that provide customized and relevant solutions, alternatives for passengers who must get to their destinations at particular times, on any airline or any other mode of transportation, instead of being booked on the specific airline's next flight due to the limitations of an e-ticket or the inability of the airline's reservation system to connect to other modes of transportation, in real time.

- Some passengers would like to have door-to-door travel with a broad spectrum of price-service options, deploying taxis, private cars, limos, small jets, and so forth. They want airlines to provide solutions by integrating with other modes of transportation in a much more comprehensive manner and using a single document (Passenger Name Record). Of course, passengers' needs and abilities to pay vary, not only by segment, but also by situation, even for the same passenger. The ability to solve an individual passenger's problem at anytime and anywhere is the key to providing personalized and customized solutions.

- As for airlines themselves, they can benefit from the deployment of truly smart, analytically-based planning systems that not only can explain what happened and what is happening, but also predict what could happen if the airline were to continue on a given path. Airlines can benefit from systems that can predict the outcome of a strategy for the entire airline as well as problem areas, by function, such as an identification that the problem is in the maintenance area, or in the operations department, or in the sales area. In fact, it is technology itself that can be used to break the silo system by forcing the systems to work from a common platform and solve the global problem from a holistic viewpoint, including indicating the resources required to solve the global problem, while reducing the planning cycle times by as much as an order of magnitude. Moreover, it is the convergence of technologies that can finally help airlines work around their institutional constraints. For example, a small segment of airlines with anti-trust immunity within an alliance can now use advanced planning systems creatively to establish "virtual" airlines that can make incredibly high margins by using information and technology to become solution providers and value integrators for small, but high-margin segments.

Management of Complexity

The airline business has been, and continues to be, extremely complex. If airlines were to pursue mass customization, were to become solution providers, and were to become integrators of value, would that not further increase complexity and, in turn, costs? Imagine the cost to gather, maintain, and synthesize huge quantities of data to develop and deliver personalized products and services. Next think of the difficulties an airline has in just coordinating its own activities across its own functions and among the alliance partners. To become a value integrator would mean coordinating with multiple external organizations. Currently, those passengers on an airline's website who wish to make hotel or car rental reservations are transferred to the websites of those companies. Imagine the cost of the complexity involved for an airline agent within its own call center to have access to the reservation system of a hotel and a car rental company at the same time to make reservations in a parallel as opposed to a sequential mode. Consider an example that is repeated in Chapter 6. A couple on leisure travel makes a booking through an airline's website or call center. Next, in making the hotel reservations, the couple discovers that arrival one day later would have saved them a significant amount of money. Going back to change the airline reservation would take more time and very likely involve change fees. If the couple were to do that anyway, the same problem could exist again when the couple moves to the next step, the car rental agency. Would it not be more valuable if all three reservations were to be made simultaneously, providing the customer with the best combination from the three companies? The standard answer from all three service providers would be that the process would be much more complex and much more costly. Although some intermediaries do offer dynamic packages, they are preset and some passengers may wish to create their own packages and have the ability to make changes in real time.

Most airlines are trying to reduce complexity, not increase it. However, although technology can be used to reduce some complexity, it can now be used to manage complexity.[6] First, creating and maintaining one data warehouse would cost less than data being collected and maintained by multiple departments. Second, all departments working with the same data and same set of assumptions (off a single platform) would lead to less analysis and more rapid responses to challenging or opportunistic developments. Third, based on Moore's Law (relating to cost, capacity, and computing power of computers), the cost of data and applications should decrease. Moreover, technology can now enable a data warehouse to become a real-time asset—an active data warehouse. This attribute involves the incorporation of new information on a passenger becoming available to all departments and at all touch points in real time.

Advancements in technology do not eliminate complexity. They can enable an airline to manage complexity in a cost-effective manner. The additional cost, if any, could be passed to the customer who might be willing to pay more for the value added. In the above scenario, for example, the passenger may be willing to pay US$50 more to the airline coordinating the package to save US$200 on the cost of hotel accommodations and car hire.

Even if complexity were to increase costs, if it also increases benefits, then the object can be to embrace and manage complexity creatively, not try to eliminate it. There are airline executives who believe that removal of complexity and the introduction of simple processes lead to a reduction in costs. That belief would appear to be reasonable if the viewpoint relates to "old complexities." Avoiding "new complexities" may not be an option as the business is becoming much more intricate with an increase in competition and an increase in ambiguity and uncertainty relating to the political and economic aspects of the global marketplace. On top of that, passengers are becoming much more demanding, empowered by relevant, timely, location-based, and actionable information. Finally, some executives are becoming stunned with data instead of intuitions from that data.

So, how can an airline manage complexity?

First is a need to replace conformist thinking with unconventional thinking. Management at one low cost airline may think that low costs can only be achieved with the use of a single aircraft type and a single cabin configuration. Management at another low cost airline appears to be quite successful with two types of aircraft and a third low cost airline with two cabins. At one low cost airline dozens of creative ideas coming from junior members of the planning team were ignored because they involved breaking the rule of single aircraft type. If adding complexity can lead to a significant enhancement in passenger experience, passengers may well be willing to pay a small premium for the better experience, not to mention a potential increase in passenger loyalty. Typically, some airlines, especially old world legacy airlines, have been stuck with an inward framework (this is how it is done in the airline industry), a historic framework (this is how we have always done it), and a command-and-control framework (this is the way the top brass insists that it be done). Airline managements could look for insights from not only best, but also next, global business practices to rethink their business models. There are examples even within the airline industry itself. Consider the ongoing thought leadership behind Air New Zealand's redesign of the economy-class seat to enhance passenger experience on long-haul flights. See the discussion and photographs in Chapter 6. Outside the airline industry, there are experience-enhancing applications on the iPhone, and the capability to capture browsing behavior by Google.

Second is the need to have the means to understand and predict passenger behavior by segment. The subtitle of this book is "accessing" passengers and "connecting" with them. While it is a fact that passengers are becoming increasingly connected, the question is, are they becoming more connected or disconnected with the airline? How many passengers buy from visiting just the website of one airline? How many websites do various types of passengers visit before they buy air travel? Would adding more functionality (for example, really advanced digital concierges) and making the website much more user friendly add to the complexity? How much more business would an airline have to receive to offset the increase in the cost for adding complexity? What would be the increase in brand equity from the increase in complexity that leads to greater value for passengers?

Third is the need to provide airline management with flexibility and agility to adapt to the marketplace that is becoming increasingly uncertain. The need to adapt to the enormous fluctuations in the price of fuel is just one example. Having multiple types of aircraft could provide one solution to combat this uncertainty. Fleet composition would entail not only aircraft of different capacities, but also age, and the method of acquisition (purchase versus lease and by type of lease). Some situations may call for the use of old fully-depreciated airplanes. Other situations may warrant the use of smaller airplanes. Yet other situations may call for the "parking" of old and fully-depreciated airplanes. Again, the cost of complexity may be lower than the value of flexibility and agility to take advantage of an opportunity.

In light of the structural and fundamental changes taking place in the global airline industry, there is strong interest in the direction of feasible airline business models. Let us start with eight basic assumptions:

- Large conventional full service network airlines are not likely to be able to reduce their costs to offer competitive service in short- and medium-haul markets to a level to compete effectively against the low cost carriers (say the International Airlines Group against easyJet and Ryanair, American against JetBlue, Air Canada against WestJet, and TAM against GOL).
- Global network carriers will continue to need traffic feed for their long-haul intercontinental flights.
- Labor at larger conventional network carriers will not agree to allow a large scale outsourcing of short-haul routes to be diverted to feeder carriers.
- Low cost carriers are likely to experience an increase in their operating costs. They pay, for example, similar prices for fuel, for new aircraft, and en-route navigational charges. And at major conventional airports they are likely to pay similar charges to those paid by the full service network carriers— JetBlue at New York's JFK and Ryanair at Madrid's Barajas Airport or Barcelona International Airport, for example.
- Low cost carriers are beginning to reach maturity relating to opportunities for further large scale growth. How much growth is left in the short-haul markets within Europe, for example, as competition increases from railroads? How many more secondary airports can Ryanair serve in Europe and still grow at its historical rate?
- Low cost carriers will not be able to duplicate efficiently and effectively anywhere near the breadth and depth of services in long-haul intercontinental markets to compete against the large full service network carriers.
- Some low cost carriers are providing service conceived to be of higher quality in short-haul markets than the large conventional network carriers (JetBlue and Virgin America in the US, for example).

- Low cost carriers have been increasing their efforts to penetrate the business market. Examples include easyJet, JetBlue, and Southwest. Virgin Blue, in fact, is reportedly dropping the "Blue" part of its name to rebrand itself to attract more business passengers.

In light of these eight assumptions, does it make sense for these two sectors within the airline industry to collaborate rather than compete for mutual benefit? The first reaction is likely to be negative. Low cost carriers have their costs low by keeping the processes simple. These include the lack of interline or code-share agreements, thus avoiding the use of sophisticated reservation systems, for example. Serving secondary airports results in lower airport costs and better turnaround times. Lack of connectivity also means there is no need for complex processes dealing with the transfer of baggage or the establishment of common standards. On the other hand, new-generation information and technologies are now becoming available for low cost airlines to interline or code-share with conventional full service network carriers without adding higher cost complexity. Moreover, the higher operating costs at major airports may be offset by the additional traffic available at major airports. Consequently, while it is not possible to totally eliminate complexity, it is possible to manage complexity to enable the two sectors of airlines to work together for the benefit of both.

Early signs are already appearing showing the feasibility of such a game-changing scenario facilitated by enabling technology. Different low cost carriers (some owned wholly or partially by conventional airlines and some independent) are beginning to develop different working relationships (interline, code-shares, and so forth) with conventional airlines. It is reported, for example, that in Europe, Vueling (owned partially by Iberia that, in turn, is now part of the International Airlines Group) feeds Iberia; in Brazil GOL feeds a number of European carriers such as Air France-KLM and Iberia and US carriers such as American and Delta; in the US, JetBlue feeds Aer Lingus, American, El Al, Emirates, and South African; and in Australia Jetstar (owned by Qantas) feeds with American and JAL; and V Australia (a division of Virgin Blue) feeds Etihad. Interestingly, Air Berlin, that had been interested in interlining with a global full service network carrier, went a step further by joining oneworld, leading to the need to interline with numerous full service network airlines within the alliance.

This game-changing business model can gain strength as the LCCs get persuaded that they can work with network carriers by managing complexity and keeping costs low enough to be in line with the revenue benefits of connecting traffic. Technology is now available, for example, relating to a newer generation of reservation systems, some upgraded from old low cost GDSs such as Navitaire and some downgraded by the old full service GDSs such as Sabre. Next, there are now similar self-service machines in operation with both sectors. Finally, there are third-party systems by Dohop, a system developed by the Icelandic technology company, to connect all the low cost carriers' offerings for a passenger through a search engine. However, since the company is already equipped with legacy carrier data, it can provide information on connections between low cost and full service network carriers. The traveler simply inputs her travel information and then the search engine provides various price-service options, involving low costs to low cost or low costs to full service. The passenger can be connected to the website of the relevant airlines. The feasibility of cooperation between the two sectors is also heightened by the fact that some low cost carriers can also make some changes to their services to enhance the overall travel experience of passengers. For example, JetBlue already provides some seats with additional legroom, a state-of-the-art terminal at New York's JFK Airport, direct TV and XM radio on-board, and reasonable snacks en route. Consequently, the services provided by the two sectors of the global airline industry could converge to a reasonable point leading to a new business model.

Capitalizing on Information and Technology

If some passengers expect airlines to become solution providers and if game-changing information and technology are available for airlines to become solution providers through the design, sale, and delivery of information-based, customized and personalized, services, and if at the same time airlines can become more flexible and agile by managing complexity, then what is the challenge for capitalizing on the new-generation information and technology? This book addresses this challenge as well as the opportunity through the next six chapters shown in Figure 1.3. These chapters take the reader through the need and the process for airlines to transform their business models by the effective deployment of new-generation information and enabling technologies with insights not only from their own successes and failures, but also from the experiences in other industries. Just keeping up with the step-changing demands of different generations of passengers is no longer an option. However, while the availability of relevant and timely information and the deployment of enabling technology can help airlines of all sizes (due to the scalability aspect of technology) to keep up with different generations of

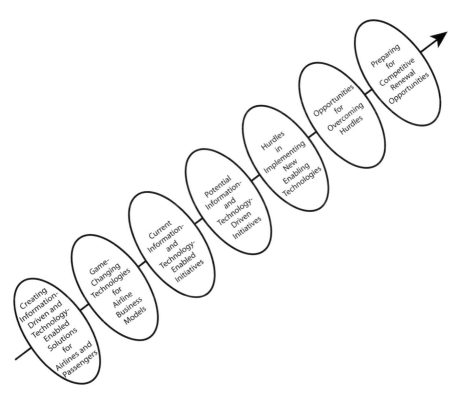

Figure 1.3　**Framework for an Airline to Become an Information-Driven Solution Provider**

passengers and develop competitive advantages, their acquisition is only part of the solution. Changing the organizational culture, structure, systems, processes and the embracement of a holistic view as well as the flawless execution of strategies are also essential to achieve customer centricity through the development and provision of new transportation solutions.

Notes

1   Joseph B. Pine II, *Mass Customization — The New Frontier in Business Competition* (Boston, MA: Harvard Business School Press, 1993) and Stan Davis, *Future Perfect, 10th anniversary edition* (Harlow, UK: Addison-Wesley, 1996).

2   "Air China value greater than United-Continental, American, JetBlue, Air Tran & US Air combined," *Centre for Asia Pacific Aviation*, March 8, 2011. The market capitalization was computed as of March 3, 2011.

3   Summarized from the book by Louis V. Gerstner Jr., *Who Says Elephants Can't Dance? : Inside IBM's Historic Turnaround* (NY: HarperCollins, 2002).

4   "The proven benefits of EFBs & ETLs," *Aircraft Commerce*, December 2010/January 2011, pp. 19–22.

5   Alaska Airlines is introducing Apple's iPad in the cockpits allowing pilots to review and update information quickly and save fuel.

6   A good description on the management of complexity is available in "Capitalizing on Complexity: Insights from the Global Chief Executive Officer Study," *Travel Industry Executive Summary*, IBM, 2010.

Chapter 2
Game-Changing Technologies for Airline Business Models

The previous chapter argued that game-changing information and technology can enable airlines to transform their business models to adapt to the fundamentally and structurally changing market environment. So, what is this game-changing technology and how is it evolving? This chapter will address the current state of the technology in five broad but inter-related areas within the Internet foundation, as well as converging technologies that can enable an airline to improve its internal planning. Whereas the first group of technologies can enable an airline to communicate and interact with passengers to improve customer service (while reducing costs and improving revenues), the second group can enable it to conduct its planning activities in an integrated framework, and at a much faster speed. Specifically, the first part of this chapter will address the five components of new-generation technology shown in Figure 2.1. These components include customized search engines, mobile applications and smartphone capabilities, context-aware and location-based mobile applications, social media and user-generated content, and intelligent virtual assistants.

The second part of this chapter will address converging and advancing technology that can enable airlines to perform integrated planning at an enormously higher pace and displaying the results of "what if" scenarios with significantly increased accuracy. Although these technologies are not new, they have been advancing (for example, computers becoming cheaper, running faster, and having more capacity, including through server virtualization)[1] and converging (for example, incorporating not only huge and integrated databases, but also business intelligence tools and applications). It is the advancements in communication technologies that provide the linkage between the Internet-based technologies discussed in the first part of this chapter and the specific airline planning technologies discussed in the second part. The next chapter will examine applications of these new-generation technologies to innovate passenger service with respect to both the airline and the airport experience.

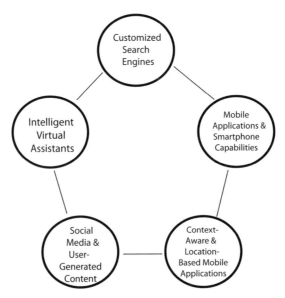

Figure 2.1   **Game-Changing Technologies Emerging from the Internet Foundation**

Harnessing the Internet: Leveraging Game-Changing Applications

Customized Search Engines

Buying an airline ticket online often involves a user starting from an airline's website, going through an OTA such as Expedia, or going through a meta-search site such as Kayak. The last option does not permit the user to purchase from the meta-search site directly. The premise behind meta-search sites is to gather and synthesize information to give travelers fare options that meet their needs and preferences. Once a selection is made, the user clicks through to the travel service provider to make the purchase. The meta-search company, in turn, collects a fee for the click through to the seller. However, it has been reported that this process is beginning to change, at least with respect to booking hotel rooms online, as Kayak began accepting hotel bookings directly on its own website as of Spring 2011. Nevertheless, meta search is still a relatively new phenomenon; in the US it is estimated that only one in five travelers have visited a meta-search engine.[2]

The differences between Web 1.0 (straightforward, one-way communication through websites) and Web 2.0 (two-way communications, content sharing, and collaborating as well as the proliferation of social networking sites such as Facebook and LinkedIn) were previously discussed in this book series. Now Web 3.0 (also referred to as the "Semantic Web") is emerging. Web 3.0 is focused on the individual and involves higher levels of personalization. Web 3.0 also relates to the method and manner in which individuals search, a development discussed later in this chapter that is shifting toward mobile browsing. In the context of a traveler, the search would combine, for example, air travel with hotel accommodations, and a car rental, with the search being conducted within an integrated framework. Consequently, from a marketing standpoint, Web 3.0 involves the marriage of new-generation technology and changing consumer buying behavior.[3] For the consumer, it provides a personalized and customized product. For the airline, it provides vital information on the buying behavior of the traveler, existing or potential.

Following is one way to describe the evolution of the Internet:

- Stage 1: One Way Information Dissemination (airline networks and services).
- Stage 2: Transaction with Passengers with Limited Interactions (reservations on airline websites).
- Stage 3: Integration with Enhanced Interaction (trip planning).
- Stage 4: Mobilization with a Higher Level of Interaction (location and time-based trip support).

This new level of information will give suppliers of travel services the ability to present newly-segmented passengers with relevant and personalized offerings. For example, in terms of search engines, a traveler should be able to conduct sophisticated, one-stop search without having to go to multiple websites. It also means the ability for a traveler to conduct an in-depth price comparison, one that truly analyzes all the possible hidden costs that could potentially drive up what initially appears to be a lower fare. It also avoids such scenarios in which a lower fare is available on days when the hotel rates are the highest. For the airline, such an evolution of the Internet means the ability to gain valuable insights about a traveler's shopping and purchasing behavior, including information regarding search behavior on the carrier's own website. If it is a potential traveler, why did he or she not purchase? Did the potential traveler simply not make a purchase, or did the traveler make a purchase on a competitor's website? Or, did the traveler end up purchasing offline? Is it even possible to collect such data, especially in terms of competitors? Even if it is possible, would the acquisition and analytical modeling of this data be cost effective? Now consider the traveler's behavior if he did make a purchase on the carrier's website. If a traveler chose the 5:30 AM flight from city A to city B, is it because the traveler really wanted that flight time, or is it because that was the only one available in his timeframe? Airlines can now implement Web analytics to be able to capture such data. The 5:30 AM flight may consistently be full, but how many passengers really wanted to be on that flight and how many took it because there was no other choice, even though this one did not meet their needs? With the current technology, it is relatively easy to capture such information in real time, literally, as the traveler makes the reservation. Optiontown is offering an interesting feature for ancillary revenue: Passengers can offer to pay a premium to get on a flight that is completely booked. The airline can then call existing passengers who are booked on lower fares to inquire if they would be willing to be incentivized to free up a seat. So by trading bookings between two or more flights, airlines and passengers can make extra money. In addition to flight time choices, there is also other valuable data that has not even been captured. A passenger could be treated very poorly on a code-shared flight but the information may be known only to the airline whose aircraft was flown during the flight. What about a passenger who made the reservation through an agent and the airline does not have the information on the passenger to contact the passenger regarding problems with the flight? Should not the airline holding the reservation call the agent and ask for contact information on each and every passenger? While this concept is not new it still has not been widely implemented due to the lack of common IT platforms.

The Internet can play an important role in building brand awareness, maintaining it, and even expanding the brand presence, and, therefore, expanding the global reach of an airline. The Internet (and, in turn, online marketing, including advertising) is now the medium to get the marketing message across. The Internet can be particularly cost-effective for small airlines whose budget would be too small for, for example, television. Traditional media may be becoming less important, but it still has a role to play. So, the mix needs to be balanced. Each carrier gets a different return on investment in each form of media. Do marketing executives have ways of measuring the return by medium? There is also the objective of the airline. Some, for example, those based in the Gulf, may be much more interested at this stage of their development in building brand awareness.

However, despite the availability of technology, many businesses, including airlines, simply have not even begun to differentiate, based on observations and synthesis of consumer behavior, as to what is relevant and what is irrelevant for a particular individual on a website. There is a lot of confusion, and, worse, frustration that online search choices create. For example, when a person inputs travel from city A to city B (500 miles away), among the search results some airlines provide ludicrous choices such as flying through a hub that is 1,200 miles in the opposite direction. Likewise, why is a possible flight pair offering that has a very tight connection time (such as 35 minutes at an airport like Chicago's O'Hare) even shown to the consumer in the search results? Airlines need to provide relevant solutions to solve passengers' needs and desires, not just be the lowest price, nor present every possible route (which just leads to confusion), no matter how much out of the way or time consuming it may be. Consumer-oriented websites are only successful if they are built around solving the problems of the consumer; they are avoided, if they obviously just push capacity into the market. In short: passengers need "buying systems," not "distribution systems."

To connect with consumers companies have used the channel of paid search that involves advertisements that companies can purchase to display when a consumer types an inquiry in the search box. However, new-generation technologies in the area of organic search can provide better results since organic search involves results that are algorithmically generated, that is, the search engines mine the Web for relevant information, extract the content, assign value relevance, and rank the pages.[4] It is, therefore, organic search that holds the key to fundamentally new opportunities and the relevant business strategies. Even in the area of paid search, new-generation technology is truly advancing the relevance of advertisements. Consider Yahoo's "SmartAds," in which one advertisement template can involve multiple variables that, in turn, can be altered in an effort to yield the most relevance to a consumer. The following example illustrates this concept. An airline carries an advertisement in the travel section of a news website. This one advertisement can be presented to different viewers based upon such factors as the travel websites the viewer has visited or even travel locations for which the viewer has purchased air travel in the past. It may also be customized to reflect such aspects as gender and particular interests. Therefore, the advertisement that a user views on his homepage in one city may reflect his interest in a particular sport and even a different price due to his geographic location, relatively higher income (for example, the purchase of a first-class airline ticket), versus the same advertisement that a user views in another city that may offer a different sport adventure and destination. Alaska Airlines is exploring this emerging technology while HP has actually implemented it.[5]

These examples provide merely a glimpse into the direction of Web 3.0 marketing, where new-generation technology is being used synergistically with the changing consumer buying behavior. Consumer behavior, in turn, is being analyzed in two ways. The first method involves collecting data (to create a profile or segment) from all the websites that a consumer views within a specific network. The second method involves collecting data based upon a consumer's behavior on one website that, in turn, may be utilized to augment the consumer's experience. This second method is widely used in e-commerce, such as in Amazon's personalized recommendations. An example of the latter method is Apple's "Just for You" feature which refines offerings to a user in real time based upon his previous purchases (whether it be songs, movies, or television shows) and based upon his ratings on suggested content.[6] Only a few airlines have begun to capitalize on technology to analyze consumer behavior at this level to convert a browser to a buyer.

Even though consumers are relatively content with online travel sites, there is still a long way to go, especially relating to the experience factor, in terms of both the site and the flight. The key is that a provider makes the data available so that a traveler may slice, dice, and aggregate data to enable the passenger to choose the best flight(s) for his needs and preferences. In order to achieve this goal, the data must be transparent. For example, flights listed as "one stop" should be further explained by "change of planes" or "same plane." This level of information might be of particular interest to a passenger traveling with small children. It is important to note that a few organizations in the travel industry have begun to do so. Furthermore, some new flight search sites have already emerged that are striving to cater to such needs and preferences of travelers. One example is Hipmunk, which differentiates itself by making all relevant information available to the user on just one page, rather than the user having to click through multiple pages. According to its website, the company strives to help users find the flight that is the right fit for them in a timely manner and without other distractions such as advertisements. The focus is on price-service options, as search results may be sorted by the following categories: arrival time, departure time, duration, price, stops, and most interestingly, what they refer to as "agony," a combination of price, duration, and number of stops.[7] Then there is Airfarewatchdog, which differentiates itself by actually using people to verify fare alerts.[8] Finally, InsideTrip differentiates itself by providing quality ratings to travelers, based upon three major components (speed, comfort, and ease) and subcomponents within each, such as number of stops, legroom, and connect time, respectively. The site even gives users the ability to calculate their own custom scores. The premise of InsideTrip's offering, as highlighted on its website, is to assist people in making smart travel purchase decisions.[9] Part of the experience factor comes down to perception versus reality. To airlines' credit, they are doing reasonably well in terms of performance, but it never seems to be good enough for some travelers. One solution may be for the search technology to go beyond internal sources and possibly even exploit cloud-based services using a contextual viewpoint (linking buyers with sellers of IT-based solutions). This technology could be particularly helpful to customer service related departments. Cloud-based services have the potential to become cost-effective for acquiring rich profile information synthesized from different sources. From a technology viewpoint this is possible. However, the concept of customer profiling information is likely to have cultural and human factor implications. Existing managers are likely to have difficulties in adapting to these concepts.

Mobile Capabilities

Mobile adoption is outpacing PC adoption. Consider also the following statistic in terms of the rate of growth: The rate of growth in smartphones has increased from one in ten in the US possessing a smartphone to one in five.[10] Finally, consider that worldwide, it is reported that there are 4.6 billion people using mobile devices, and 1.7 billion using the Internet.[11] What factors lie behind this explosive growth in mobile? Some factors that have accelerated the growth in mobile include:[12]

- the available financing through venture capital of mobile applications
- the increase in service coverage
- the increase in data infrastructure (reaching 3G/4G)
- the lowering of cost barriers in terms of hardware, service, software
- the technological advances in smartphones
- the integration of a wide array of GPS-based applications such as navigation, location-based services, etcetera
- the movement of consumers away from their PCs as they are always on the go and have a need for mobility

The demand for mobile devices will continue to grow given the almost unlimited low cost bandwidth (so a user can attach pictures like seat maps), low cost of communications, and the enormous processing power of the units at relatively low costs and in real time. Consider, for example, a few advancements in smartphones, such as by Apple, the leader in the mobile arena (with the introduction of its iPhone 4). Key features of the iPhone 4, as highlighted on the company's website, include:

- FaceTime (the ability to talk face-to-face with another iPhone 4 user over Wi-Fi)
- Multitasking (the ability to run third-party applications and switch back and forth between them, all without the concern of slowing down the system or draining the battery)
- HD Video Recording (the ability to shoot video, edit it, and even share it with others all from the iPhone 4)[13]

The concept of mobility continues to evolve with the emergence of offerings such as the iPad by Apple. The iPad allows a user to consume a myriad of types of media (books, movies, magazines, newspapers, video games, music videos) practically anywhere, anytime. It also allows a user to have access to functionalities such as a calendar, contacts, and the transmission of photos. It even has the potential to be a game changer in that the device may just change personal computing as it is known today. For example, while an iPad user is able to run most of the 150,000 applications that are available on the iPhone and the iPod touch, there are also three applications that Apple is featuring with the iPad that illustrate the business potential of the device. These applications include (1) Pages, for word processing, (2) Numbers, for spreadsheets, and (3) Keynote, for presentations (all three are rewritten versions of the Mac's iWork office software suite). Another feature that is available in conjunction with the iPad that makes it a possible rival with the laptop is a new combination of a keyboard/docking station that positions the iPad at an optimal angle for typing.[14] Finally, a large screen allows users to view full pages (in portrait or landscape) easily and in amazing clarity, as well as the utilization of an expansive on-screen keyboard by merely touching the screen. The device is lightweight and easy to handle. The iPad is sleek and easy-to-use, in alignment with all of Apple's offerings. Just as the trend has evolved from PCs to laptops to mobile phones to smartphones, smart tablets (such as the iPad) or tablet PCs can now be added to the list of emerging and evolving technology, which, in turn, provides users with increasing ease of use as well as functionality.

Following are a few business examples of how companies, airline and non-airline, are implementing the iPad into their business strategies. Hyundai offered an iPad to each customer who purchased the automaker's new luxury offering, the Equus. The idea behind the iPad was twofold. First, it replaced the owner's traditional paper manual. Second, it had the capability to enable users to schedule service appointments online as well as to make arrangements for a pickup to take the car in for maintenance.[15] Mercedes-Benz also incorporated the iPad into its business, but in a different manner. Mercedes-Benz Financial wanted to equip a number of its dealerships with an iPad. The logic behind the pilot program is that customers stay close to the product during the purchase process, rather than sitting in a cubicle or in an office. An application on the iPad allows dealerships to conduct business right at the location of the vehicle, such as checking on available incentives (via the Vehicle Identification Number), initiating the credit application process, or retrieving information on an existing customer. Dealerships may also use the iPad to enter notes regarding the condition of a returned lease vehicle.[16] Within the airline industry, Malaysia Airlines announced its launch of kiosks selling airline tickets using the Apple iPad in Kuala Lumpur, enabling the carrier's passengers to book, pay, and check-in through the iPad kiosks.[17] Finnair launched a program in which complimentary iPads were distributed on some flights to and from Hong Kong in both business and economy classes for entertainment purposes but also for customer feedback in terms of potential product and service offerings.[18] Moreover, some airlines are considering replacing their existing expensive and complex in-flight entertainment systems with easy-to-use iPads. One example of an airline that is moving toward this potential trend is Jetstar, who is experimenting with it as a rental device for in-flight entertainment. Internal processes of airlines in many cases involve mobile employees who could become much more productive using iPad-like devices inside a terminal, out on the ramp, or inside a maintenance hangar.

Mobile is a distinctive channel that can transform the manner in which passengers research, buy, consume (especially due to the proliferation of applications), and share their travel experiences (especially due to the proliferation of social media). One means by which mobile can offer an enhanced experience is by enabling a user to obtain additional information on an object by merely pointing her mobile device at the object, for example the Eiffel Tower in France, the Great Wall of China, or the Taj Mahal in India. One example of this type of mobile offering is Google's "Google Goggles," that enables a user to use her pictures taken with her mobile phone to search the Web. A user just accesses the application, takes a picture, and then the search results appear. The key feature being that the user does not have to type in a query.[19] Consider, for example, a business case of mobile in terms of capturing objects while out and about, the American paint company Benjamin Moore. Specifically, the company has created a mobile application (the Color Capture iPhone application) that allows a user to point at anything (sunset, grass, building), take a picture of it, and then the application will provide the closest matching colors that the company offers in its catalog, including the coordination of colors. The application also provides users with directions to Benjamin Moore retailers in the area.[20]

Despite the fact that some executives continue to question the impact of the mobile device due to some of their limitations (such as small screens and ease of navigation) mobile is predicted to have much more impact than the Internet, as the mobile device is the key to connecting with customers on a personalized and individual level. Consider this point. If companies were concerned that potential customers might go online and read a negative review about their companies or see a search result that led to their competitors, think about the impact of having customers literally stand in a store and use their mobile devices to (1) price shop, and (2) locate the nearest competitor.[21] One example is Amazon, who launched a price check application for the Apple iPhone (and also the Apple iPod touch) in the US, which gives consumers the ability to compare the price of a product in a store with that of Amazon and other online retailers by any of the following ways: (1) scan the bar code of the product, (2) take a picture of the product, (3) voice search, or (4) text search.[22]

Some mobile-savvy businesses are actually using the channel to promote such activities (described above) by consumers. When global cosmetics retailer Sephora first launched a mobile site, it mainly focused upon ratings and reviews. However, in mid-2010, Sephora North America took this concept to a new level with the launch of its mobile commerce site. Specifically, the new offering, which may be accessed through all smartphones, (1) provides information regarding products, (2) enables users to search and buy products, (3) enables users to create a shopping list, (4) incorporates its loyalty program, and (5) equips users with their purchase history. The last feature is particularly customer centric as it helps a consumer if she cannot remember the exact brand or color of product that she bought in the past. Finally, Sephora North America launched a mobile application (available for the iPhone) that enables users to access special features such as the ability to scan the bar code of a product and instantly obtain information as well as access to "How To" videos.[23]

What could be some meaningful practices for airlines in terms of mobile, incorporating the advancing computing functionality to provide information-based customized services? One opportunity is the potential use of mobile devices to transmit data to a passenger regarding what the *true* effect of a 30-minute flight delay could mean in terms of the final arrival time. A mere 30-minute flight delay due to a late inbound flight could translate into an actual 3-hour delay for a passenger due to additional delays relating to connections, airport congestion, and so forth, including the possibility of spending the night at the location of the delay. Mobile devices could be used to communicate all of these additional potential delays so that a passenger may be equipped with the accurate data to make appropriate alternate travel plans at the time of the announced delay, rather than waiting for hours only to be informed that the flight has, in fact, been cancelled. Technology is available to process this information for each and every affected passenger and provide a personalized and customized solution in real time. Some passengers would not only be willing to pay for such information but loyalty could be taken to unheard of levels. Also, the owners of mis-connected baggage could be informed proactively, saving time and frustration when waiting at the baggage claim in vain. Instead passengers could communicate the address where they expect the delayed baggage to be delivered (home, hotel, resort, friends' house, and so on). For more on mis-connected baggage management through leveraging new-generation information and technology, see Chapter 3. Furthermore, for information on mis-connected passengers, see the example in the Appendix.

As for the limitation of the small size of the screen, it has been reported that many people actually enjoy watching TV on mobile devices. For example, South Koreans are actually the leaders in terms of viewing television via mobile devices; an astonishing 27 million people, equating to 56 percent of the country, watch mobile TV on a regular basis. Mobile technology is also becoming prevalent in Africa, China, India, Latin America, and Southeast Asia, where it is estimated that 80 million people now possess mobile devices that have the ability to receive free, live television broadcasts. Interestingly, this phenomena is not the case in the US and Europe, attributable, perhaps, to operator resistance, discrepancies in technical standards, and licensing hurdles, all of which have limited the technology.[24]

Context-Aware Computing is taking off as a result of expanding cost-effective wireless technologies, increasing power and intelligence of mobile devices, and advancements in predictive analytical tools. The result is the capability to integrate text, data, graphics, audio, video and so forth and provide the information within the framework of a context (to an individual person at a unique location, in a specific language, and relating to a specific situation). This capability is important as passengers are demanding not only faster and much more complex responses, but also customized responses.

Mobile Applications

Mobile applications refer to programs that may be added to mobile phones to personalize the device to the users' needs and interests. These applications can consist of a myriad of different types of programs including entertainment (such as games), information (such as maps, news, transportation schedules, and weather), services (such as banking), shopping (such as Amazon and eBay), social networking (such as Facebook and LinkedIn), travel (such as airlines, hotels), and so forth. Different mobile applications are available for different mobile devices. Users can download or purchase an application from a variety of sources. For example, there is Apple's App Store, as well as Google's Android Market, Research in Motion's (creators of the BlackBerry) App World, and there is also the Windows Mobile Catalog. Some applications are available at large on the Web, although not associated with any particular store.[25]

Just as "smart card" technology has been implemented to serve as payment systems and maintain consumer information, mobile devices can be utilized in the same manner. Instead of using a "smart card" to gain access to a hotel room, a bar code could be sent to a mobile device for the user to gain access to her hotel room, just as a bar code may be used as a boarding pass. Within airports, the device could be used to allow a much faster passenger handling than today's bar-code readers. As in the case of a "smart card," mobile devices may be used as a form of payment when checking out at a store. Such capabilities can be easily incorporated into smartphones through the use of RFID chips to make smartphones act as "smart cards." (For more on RFID, see Chapter 3).

Smart companies are capitalizing on the mobile channel by launching company-specific applications to cater to their busy customers who need to conduct business on the go. One example is FedEx Mobile for iPhone, that can enable a user to not only track the status of a package, but even create a shipping label right from the hand-held device. A user may also locate the closest drop-off center or even request a pick-up from the device.[26] Similarly, megabus.com (described in Chapter 4) offers pick-up points right on the street which travelers can find using their mobile devices, rather than having to deal with terminals. Consider the financial services industry. USAA, a bank and insurance company focused on military members and their families, has capitalized on the fact that many of its clients have unique needs given that they are very mobile. In an effort to cater to their needs, the company has leveraged technology to implement customized offerings, such as the ability of its customers to make iPhone deposits (a customer takes a picture of the check with her iPhone, then uses an application to send it to her bank, and the money is available in her account within minutes) and to receive texts with balance information, while in the field. The company also launched a similar mobile service deposit application for Android users. Other programs include plans to offer a mobile peer-to-peer payment system, in which customers may email or text money to friends or relatives for immediate deposit, regardless of their locations, as well as a mobile car purchasing service in which a customer in a dealership can take a picture of the Vehicle Identification Number on a car via his iPhone and then receive insurance quotes and loan terms instantly.[27]An insurance company in Austria offers skiing insurance that can be bought via a mobile device right while riding in the ski lift, which is certainly a great example of a company capitalizing on the mobile channel's anywhere, anytime advantage. Similarly, airlines can capitalize on mobile technology to offer last minute empty seats to passengers who wish to have extra space while traveling (for more on this concept, see Chapter 4).

Mobile Applications versus Mobile Websites

While many features of mobile applications and mobile websites are very similar, and thus may result in a similar experience for the user, it is important to distinguish the differences between the two mediums, and to determine which one is most appropriate for a particular business. First, normally, a mobile application only runs on the device for which it was built, such as the iPhone. However, a mobile website can be used on any mobile device that has Web capabilities. Therefore, mobile applications may be more appropriate for niche markets, whereas mobile websites may be a better fit when a business is trying to reach out to a larger audience. Since mobile applications can work together with a particular device's internal functions, they work well for specific or complex requirements. One example would be a banking application, where a customer could locate an ATM utilizing the GPS specific to the device. This feature would yield a better experience for the consumer than going to a mobile website and having to type in the zip code and scrolling through several screens to obtain the desired information. Mobile applications can take advantage of other features specific to the phone such as the camera. Furthermore, applications have the ability to work in conjunction with other services such as Facebook. However, mobile websites may be more appropriate for providing content or for shopping as the content is accessible by search engines.[28]

Context-Aware and Location-Based Mobile Applications

If we assume that two key areas of step-changing technology relate to digital and mobile, what does the shift to mobile mean for businesses, in our case, airlines? It means that they need to change the way they communicate with consumers. In the case of airlines, mobile devices allow passengers to gain access to almost anything, anywhere, anytime. This level of global connectivity gives travel service providers a great opportunity to, theoretically, truly access and cater to an individual passenger anywhere, thus giving the potential to become more customer centric. It also means that the bar has been raised; consumers expect more. While some other businesses have had success meeting this level of expectation, airlines have, in general, fallen short thus far. One good example of a service provider that has leveraged mobile technology is restaurants. Rather than using traditional pagers to inform customers when their table is ready, some restaurants are implementing technology that allows them to send a text message to the customer's mobile device instead. From an operational standpoint, this use of technology is positive for the customer as it eliminates the constraint of remaining within the restaurant, and it is also efficient for the restaurant as it does not have to be concerned with maintaining the pagers. From a marketing standpoint, the restaurant could take this concept one step further: enhance the customer experience through this process by offering vouchers, sent directly to the mobile device, for a complimentary beverage or appetizer, if the quoted wait time is longer than originally promised. While the restaurant industry does not face many of the burdens and constraints that are inherent to the airline industry, this example does provide insight into how a service business may leverage information and technology to move toward becoming more customer centric. For more on technology in the restaurant industry, see case study on OpenTable in Chapter 3.

A myriad of applications are being launched based upon the location functionality. Specifically, the combination of location-aware devices and applications is unveiling information that consumers desire, everything from "Trapster," which provides information regarding the location of police traps (drivers report red light cameras and speed cameras as well as the location where police are hiding, all of which are added to the map of police traps), to "SitOrSquat," which provides public toilets in the user's vicinity, as well as user ratings on the facilities. Both of these applications are free. The combination of location-aware devices and applications is also unleashing a variety of service-oriented applications, such as "iNap," which provides a wake up alarm so the user does not miss her stop while on public transit. This service, which does have a cost associated with it, allows a user to indicate where she will be disembarking via a Google map in the application. The application then tracks her location and alarms the user before arriving at her destination.[29] The German railroads offer an application called "Railnavigator." One feature is a "take-me-home" feature, which suggests the next train to get to your hometown, depending on your present time and location. It also guides the user to the nearest entry point into the train system and allows one to purchase a ticket "on the go." Similarly, in Switzerland, users can deploy an application called "GottaGo" to help them determine the best option to get where they need to go (home, work, and so on) in terms of public transportation.

Consider the location-based mobile application, Foursquare, that allows members to meet up with friends and to discover new places. Members can earn badges by digitally checking in at physical locations including bars, restaurants, and other venues. Checking in involves clicking a button on the application when entering the venue. When a user checks in, the user's mobile device determines his or her location, and then conveys this information to those that the user has chosen. The badges then become part of the member's profile. Foursquare's business model involves creating a large user pool and then selling businesses on the opportunity to market to such a group. This concept is similar to the opportunity that the airlines could have when Wi-Fi becomes a standard amenity on-board: airlines could sell businesses on the opportunity to market to such a captive audience. What is next for Foursquare? It has been reported that the company is working toward the goal of enabling the following scenario: at the end of the workday, a user's phone informs her of her plans for the evening, after the phone has already checked her friends' calendars as well as has checked if the restaurant they want to try has a table available. Finally, the phone lets the user know that a few friends will be at a nearby venue so that they can all meet up later in the evening.[30]

Applications are, of course, a key part in the mobile technology field. Take, for example, the University of Maryland, that now offers courses on the topic. The university features a course, taught by a visiting Apple engineer, in which students have the opportunity to work with cutting-edge technology. Students learn how to create basic applications, such as those that help individuals get around the campus or find dining establishments.[31]

Moreover, Apple's future iPhone application, iTravel, could prove to be a game changer in the travel industry as the application promises to go beyond the elimination of paper tickets and self-check-in capabilities to services such as baggage identification, car rentals, hotel reservations, as well as other travel services, while on the go and within a personalized framework. The idea being that the iPhone would interact directly with airline kiosks, baggage systems, boarding gates and eventually security checkpoints.

Think about the massive potential opportunities that such a concept presents. However, it is also important to note that there are some concerns regarding such location-based applications. First, even the largest of such networks consists of only a few million members, and thus has a far smaller reach than other mobile marketing channels such as SMS and mobile search advertisements. Second, there is the issue of individuals tampering with the system, such as finding ways of checking into venues when in fact they have not actually been there. Third, there is the privacy issue; networks will have to balance sharing information about users' whereabouts with potential partners versus potential threats, such as robberies.[32]

Social Media and User-Generated Content

Social media is certainly gaining momentum. Consider the fact that it only took Facebook to add 100 million users less than 9 months; to reach 50 million users, it took the iPod 3 years, the Internet 4 years, the TV 13 years, and the radio 38 years.[33] While a few airlines are capitalizing on this technology, some airline managements continue to question as to if it is a fad. It is true that some social networking sites become extremely popular, only to then slowly fade out of the limelight. However, Facebook, at the beginning of 2010, was reported to be the second most popular site on the Internet after Google and appears to be the social networking site of choice and not just among Americans. It has been reported that as of early 2010, some 70 percent of its audience is outside the US.[34] One interesting observation about social networks in general is that they have been able to adapt to cross cultures at a rapid pace. Although there are some differences in the local usage of social sites (examples include higher use of social sites to play online games in some regions and dating services in some other regions), services such as Facebook and Twitter have become a truly global phenomenon. Regardless of the minor differences in use of the social sites across the globe, the new-generation technology will take the general use of social sites to new heights. For example, social sites are now becoming mobile. Furthermore, technology can now enable the content to contain not just pictures but also audio, video, streaming media, and augmented reality browsers.

A study by Nielsen conducted in October 2009 indicated that Australia was the leading country in terms of average time spent on social networking sites, followed by Britain and Italy, with the US ranking fourth.[35] Therefore airlines cannot overlook the power of these communities, especially not Facebook. If its gigantic membership base started to buy on a wholesale collective basis and sell on a retail basis, the impact would be profound. The issue is not simply to ignore such communities but rather how to interact with them to turn the challenge into an opportunity in extracting intelligence from information. Airlines may view social networks as either a challenge or an opportunity. Take for example, the 500 million Facebook users. Suppose each member was to indicate his or her travel plans for the next six months. Technology could easily be used to segment this data in numerous ways such as by day and by market. Facebook could then use this information to obtain competitive bids from different airlines for the services needed by the Facebook users. On the other hand, airlines could work proactively with Facebook to identify and provide value propositions to the Facebook members, thereby coupling social network and data mining technologies.

Some companies jump quickly into the world of social media, often being taken aback by the complexity and speed of the medium. However, a holistic strategy is needed when venturing into social media. Otherwise, fragmented approaches can yield negative results, including incomplete data, inconsistent experiences, and the lack of ability to collect and collate consistent feedback from the customer across all touch points of the organization. There are several key points to consider when launching a social media strategy. First, the business objective(s) that the social media is expected to support should be identified. Once the objective is defined, it may be determined how social media can be leveraged to help meet this goal. Next, the metrics that social media is expected to improve should be identified. Finally, the real-time aspect and speed aspects associated with social media must be considered when establishing policies, procedures, as well as the people and culture of an organization.[36]

Measuring social media is a unique challenge because not only does it involve the interactions between potential consumers and an organization as well as current consumers and organization, but it also involves interactions between consumers and other consumers, and even those between consumers and other organizations. Furthermore, much of the data involved in social media is unstructured (such as conversations) where as other interactions through more traditional mediums (such as customer surveys) are of a more structured nature, involving well thought out questions and offering multiple choice answers. Finally, it is sometimes difficult to determine where the customer is in terms of his relationship with the company, who the customer is, or for that matter, if he even is a customer. Again, this is unlike traditional interactions, such as when a person places a call to a customer support center, and the company can identify if the person is an existing customer, details of his purchases, accounts, and so on. Despite these challenges, there is much to gain in terms of obtaining insights into consumers with respect to their feelings about a company, future behavior predictions, as well as consideration for customers' unmet needs. There are a few predictors that can help assess these sentiments. These include the feedback that a firm receives from customers, the information that customers share with other consumers, how customers interact with other organizations (whether it be competitors or partners), and how a customer interacts with a firm. These predictors represent the "health" of customer relationships. The merging of such new data with traditional CRM data can help companies gain a full view of its customer.[37]

As companies move from experimenting with social media (such as implementing social media tools) to integrating it into the organization's operations (strategic holistic customer relationship management), it is necessary to implement a social media technology platform. The idea is to identify those conversations about the company and direct the data to the appropriate people within the organization. Specifically, the platform receives input from various social media tools, and then implements analytics software to monitor trends. The key is to link the social media platform with the organization's current CRM system. This can be a hurdle as it involves combining unstructured data involved with social media and structured data associated with traditional CRM. However, doing so could dramatically help organizations in gaining a holistic view of consumers.[38]

One company that has incorporated social media as part of its business model is US-based Groupon. Groupon leverages what they refer to as "collective buying power" to bring local businesses new clients and local consumers deals on experiences in a particular market. It is easy for consumers to use: they simply sign up to receive the daily deal emails for their local market. They may also receive the information through Facebook or Twitter. The deals include a myriad of goods and services such as deals on restaurants, spa services, dry cleaning services, lessons for everything from horseback riding to flying, movie tickets, and even travel services such as bus tours. Groupon negotiates large discounts with local businesses, but a certain number of consumers must sign up for the deal to go through. The company offers these businesses a means to guarantee a return on their promotion budget, thus obtaining access to a large advertising pool.[39] Users are encouraged to share the daily deal via email and promote the deal through their social networks in an effort to increase the number of consumers that sign up for the deal, and ultimately, activate the deal. Here is how it works: On a given day, a user checks out the daily deal. If she is interested, she enters her name and credit card number. However, the credit card is only charged if the deal is activated (if enough consumers sign up for that particular deal). If the deal is in fact activated, the user is sent a voucher, which she may use by printing it or presenting it via her mobile device, just like cash. The company even offers a promise to its patrons, as highlighted on its website, illustrating its commitment to treating its customers well. Specifically, if customers are unhappy, they are encouraged to call the company, who offers to return the purchase.[40] Groupon has exploded from a start-up to a rapidly growing profitable business. What does such a business model mean for the airlines? For one, they could charter a plane when there are enough members willing to do the deal. Hence, social networks can provide significant opportunities for those airlines that use the medium proactively.

A number of competitors have embraced the group buying concept and have launched similar offerings including BuyWithMe and LivingSocial. Moreover, a collective buying site has also emerged within the travel industry, TripAlertz, which features discounted vacations. Here is how it works: members "vote" on presented deals. Deals with the most "votes" are then offered as the travel deal. Travel deals are offered for two weeks (unlike Groupon's daily deal). While travelers book at the same time, they do not have to travel together. While members will always receive the discounted member price for the vacation, they could pay even less; the more members that book a deal, the lower the cost of the vacation.

Forward-thinking businesses are not only connecting with their customers via social media, but they are recognizing that social networking is also becoming mobile. Let us start with a company from outside of the airline industry, Unilever, which clearly recognizes the power of social media in terms of building brand ambassadors (as described by Laura Klauberg, Senior Vice President of Global Media at Unilever).[41] As she points out, one key advantage of digital marketing initiatives is that they provide an opportunity to engage with customers as opposed to the one-way communication through television and print media. However, businesses must go beyond the switch from the use of television and print media to digital media. The key is to define how consumers use digital media and then develop marketing initiatives that develop the brand around the way consumers interact and engage with the brand. Unilever developed its Campaign for Real Beauty for its Dove brand in which it provides an opportunity for women to speak out about a subject for which they feel passionate, in this case, beauty. In this case, social media is an ideal platform for women to communicate their messages (posting videos on their Facebook pages or MySpace profiles, or writing on the Dove message board). In essence, through these digital channels, these consumers are creating advertising for the brand. Although, Unilever has always promoted Dove with the use of the testimonial framework using real women, it is the creative use of the digital medium to engage and interact with consumers on the subject matter that is relevant and important to consumers, and that has taken the brand to new heights.

American fast food chain Pizza Hut appears to be capitalizing on both social networking and mobility. The company launched a Facebook application which allows users to order pizza without leaving their profile pages. Not only does such a tool increase sales, but it also allows the company to capture individual customer data that would not be possible through phone orders or paper coupons. The company also has launched an iPhone application that enables users to place an order while on the go.[42]

Tired of inaccurate and untimely information, some travelers have taken the initiative to create their own travel information websites. One example is FlightCaster.com, which was launched in an effort to assist the frequent traveler by frequent fellow travelers who have grown weary of being reliant on airlines to provide flight status information. FlightCaster recognizes that information is power in travel, as accurate information results in more options as well as possible solutions in the case of delays or cancellations. The service predicts flight delays (currently in US domestic markets), using an advanced algorithm that analyzes data on past domestic flights and applies it to real-time conditions.[43]

A traveler, frustrated by the lack of information available to passengers while in airports, started a mobile application, GateGuru. The application (currently available for the iPhone) allows a user to find eateries as well as shopping and services in his vicinity within an airport. Most importantly, the application lets a user know what is available near each gate so the traveler knows what options are available after going through security. In addition to providing dining, shopping, and service information, the application gives ratings and reviews of the offerings by other travelers, in an attempt to prevent a user from having to eat a terrible meal, or worse, go without. Photos are even available through the application. The service is currently available for numerous US airports, as well as in Canada and at London Heathrow. GateGuru, as highlighted on its website, is about leveraging technology and empowering travelers, all in an attempt to augment the customer experience. It takes the unfamiliar and turns it into a positive experience for a traveler in the airport environment.[44]

HasWifi.com is another website, established by an individual who was frustrated by the fact that he was unable to do any online work on a number of flights that he had taken. The purpose of the site is to enable travelers to plan ahead on whether or not their upcoming flights offer Wi-Fi on-board. The founder of the service understands that many travelers wish and need to be connected at all times, and it gives them this information in a quick and easy manner. Users simply select their carrier and input their flight number on the website. The site allows travelers to avoid the disappointment of finding out that their flight does not have Wi-Fi as they will know before they even book their flight. The service is powered by a combination of data mining as well as user contribution. Specifically, the "You Tell Us" feature on HasWifi.com allows users to provide feedback as to whether their recent flight offered Internet on-board.[45]

SeatGuru.com was launched by a frequent flyer who had experienced the vast differences among airplane seats and wished to compile this information in an effort to share it with other travelers. The site offers color-coded interactive seating charts for hundreds of airplanes and a myriad of airlines. The color codes indicate good seats, standard seats, and poor seats. The location of galleys, lavatories, and closets are also noted. Finally, on each airplane page, there is an In-Flight Amenity box that contains icons which indicate the services available, such as audio, video, power outlets, Internet and even infant accommodations (such as bassinets), on a particular aircraft. A user may click on each icon for additional details and even links to further information. SeatGuru was purchased by TripAdvisor.com in 2007. SeatGuru also relies on user input in terms of feedback on seat width, pitch, and comfort.[46]

Intelligent Virtual Assistants

Virtual assistants are being introduced in a variety of areas due to advances in technology. These intelligent virtual assistants, coupled with customized search engines, hold a significant potential to finally providing personalized passenger service. For example, technology in terms of speech recognition has progressed to the point that, in the case of the airline industry, if a passenger places a call to an airline, the airline can have the technology in place to ascertain the gravity of the situation by detecting the traveler's emotions.

Think about the level of sophistication that can be achieved by the use of fundamentally step-changing technology, illustrated by the development of the Watson supercomputer by IBM. Watson is a system that can answer questions formed in natural language (the manner in which humans actually communicate) through the use of unstructured text analytics to comprehend the question, analyze vast amounts of information, and give the best response. One recent example (during February 2011) of Watson is on the US game show *Jeopardy* in which Watson competed against the two top human contestants and won. In this role, Watson was required to not only deploy encyclopedic knowledge, but also the ability to understand rapidly complex and unclear statements in the natural language. The core objective of the system appears to be to communicate and collaborate on human terms through natural language.[47] The concept clearly has an application to the airline industry in terms of their call centers that handle millions of calls. The call center hurdles are primarily related to cost and wait times. Implementing such technology can enable an airline to increase service quality through the provision of rapid, efficient responses to customers. Moreover, such technology can allow people to capture the data that is required in the manner that they need, not just how it is structured and presented through the system.

Given these advances, an airline's system could detect the level of urgency, and therefore provide the appropriate depth and breadth of solutions to the traveler. According to one hospitality study, hotel rooms of the future may have televisions which utilize voice recognition to answer guest inquiries in an attempt to eliminate the need for guests to call the front desk with inquiries.[48] Technology is also being leveraged to enable airlines to augment their websites with intelligent virtual travel assistants, designed to answer questions and direct travelers to the information that they are seeking on the carriers' websites. (For more regarding airline websites and virtual assistants, see Chapter 3.)

While currently travelers type their question and the virtual assistant responds via voice, could airlines implement such technology to enable potential travelers to ask (via the voice form) questions while on the website? Is there a back-up flight in case I miss the connection at the hub? Is this first-class seat on the Chicago–Honolulu flight simply a standard first-class seat in US domestic markets that reclines a couple of inches or is it a lie-flat seat? I do not hold a membership in your airport lounges, but will this first-class seat between Chicago and Honolulu allow me to use your airport lounge? Why is it that a business-class seat between Chicago and Heathrow will allow the use of the lounge but a first-class seat between Chicago and Los Angeles will not? On the other side, the virtual assistant could step forward and ask the traveler if he was aware of the fact that the flight connection the passenger selected has a six-hour layover or that a 55-minute connection time at JFK at that hour is unrealistic. Such technology could allow for a totally interactive interface between the airline and the traveler, thus truly augmenting the person's experience with the airline even before the travel commences.

The two-way interaction described above would certainly be a far more passenger-centric method than the current way in which airlines have developed complex IT self-service portals where the object is for a caller to get his questions answered through a series of options. Supposedly, this saves an airline costs. Does it? Most issues still require support from a person, meaning that not only is the cost still there, but the passenger's frustration level has been raised to an unnecessarily high level. What is needed is a combination of virtual assistants and knowledge management technology applied to a carefully segmented demand of calls. Are there any surveys of how satisfied or dissatisfied callers are with the IT self-service phone options? Just as important, is data gathered and analyzed on the questions asked by the callers? Is such data passed to the appropriate departments?

Will businesses deploy not only advances in individual components of technology, but also the technology that connects the different components? One thing is for certain. It will be consumers, not businesses, which will determine what upcoming technologies are needed and deployed. One potential example from the hotel industry is the concept of a pillow equipped with built-in speakers and wireless capability to cater to those who need to conduct conference calls.[49] What does the airline industry look like in the totally connected world?

Another point, smartphones can only carry a limited number of applications. So there will be a competition for that scarce space. All airlines, hotel chains, rental car companies, meta-search engines, airport operators, and so forth cannot make it to the pocket of the consumer. The interesting question will be whether a "meta-travel application" (such as Apple's iTravel) will become the integrator of all other travel related applications, thus eventually gaining ownership of the customer.

Integrated and Close to Real-Time Airline Planning

While the previous section focused on new-generation technologies that centered on the marketing and delivery of products and services, this section provides a broad brush view of advancing technologies for airlines to improve their internal planning activities in three areas:

- First, all planning initiatives can now be fully *integrated* in that all functions use the same data and the same assumptions, and that all activities can now be performed in a simultaneous manner, as opposed to a sequential manner. The integration can be between the commercial and operations areas, for example, in the scheduling of aircraft, crews, maintenance, and ground operations. Or, it can be totally within one area such as commercial (as shown in Figure 2.2).
- Second, planning scenarios can now be evaluated in an order of magnitude of less time, in days as opposed to months. These advancing and converging technologies can now provide robust answers to "What if" questions such as: "What would we do if we had 20 more aircraft?"; "What would we do if the price of fuel went up to US$120?" Answers would be produced that are not only fast, but also much more robust.
- Third, while complexity cannot be eliminated, advancing and converging technologies can enable an airline to better manage complexity. There are two key benefits relating to better management of complexity. First, the complexity that an airline does decide to keep (such as a large number of aircraft in the fleet) can be maintained at a lower cost. Second, the complexity that an airline decides to keep to benefit the passenger (for example, a much richer mix of price-service options) can be managed in the "back office," simplifying the passenger's decision-making process.

Let us start with network and scheduling, functions that are not only the heart and soul of an airline, they are also extremely complex, involve huge and disparate databases, and subject to numerous changes. High-speed and high-capacity computers (using business intelligence tools and applications, service-oriented architectures, enterprise resource planning systems, metadata systems, user preference analytics, interactive dashboards, and so forth) can now forecast more effectively the projected performance of changes in an airline's network and schedules and to analyze the actual post-departure performance by flight and by cause of variance between the forecast and actual performance. For example, poor profitability on an individual segment can be caused not only by numerous factors relating to demand (further divided by point-to-point versus connecting traffic, intra- versus inter-line traffic), load factors, fare mix, and poor scheduling, but it can also explain inter-relationships. Taking this example one step deeper, connecting traffic may not only be the cause of the lower profitability, but it may also be the cause of some flight delays that, in turn, have a domino effect. There are a number of key points regarding the use of advancing and converging technologies:

- First, technology is now available for not only creating an *integrated* schedule a year or two ahead, but also for managing it as unforeseen events take place closer to and including on the day of operation (again, from an *integrated* and *holistic* viewpoint).New technology can enable true integration even within the existence of the silo system (for example, between aircraft, maintenance, and crew scheduling/rostering, ground services, and, in turn, with pricing, revenue management, and sales).
- Second, technology can now free the analyst from working with averages on load factors, fares, costs, and so forth. It is now possible to drill down to the lowest level to determine the cause and effect (all flights in a market or an individual flight, actual fares and fare mixes versus averages, low-yield connecting traffic that is coming from a particular point of sale or a particular alliance partner on a particular day, beyond revenue by flight, by destination, and by direction).
- Airline disruptions are commonly associated with deviances from the originally planned operations caused by external events (such as ATC congestion) or internal occurrences (such as unavailability of a gate or the late arrival of the catering truck). The problem at most airlines is that information is kept only at the macro level, such as delays measured by government standards. Technology is now available to dig deep into the core causes and the multidimensional effects of these airline disruptions, leading ultimately to not only savings in costs but also improvements in service. Interestingly, there is even a connection between the airline-

specific technologies and the Web-based applications. Web browser and Web applications can be used to integrate with multidimensional but disparate data sources, through the Internet or an airline's intranet. The challenge is to blend the interests of all stakeholders.

- The turnaround time for the analysis can now be so short that the results of the analysis can lead to the deployment of effective responses in a heretofore unbelievable time framework. Budgeted costs can be compared with actual costs almost in a real-time framework and updated accordingly for the upcoming flights.

Next, consider different sub-functions (shown in Figure 2.2) within the commercial area. While airlines have used older-generation technology to analyze fares and fare classes as well as the performance of revenue management, advancing and converging technologies are now becoming available (for example, user preference analytics and advanced communication capability between an airline and its customers):

- to integrate the traditional revenue-related functions (pricing structure and levels, revenue management, sales, and so forth) among themselves as well as among the different distribution channels
- to integrate the new revenue function, ancillary revenue, with the aforementioned revenue-related functions, customer relationship management, loyalty, and customer service
- to provide "plugs" at various revenue leakage points, both internal (charges for extra baggage, changes in reservations, no-shows, and so forth) and external (ticketing by various types of agents according to rules, bad bookings, and so forth) to achieve revenue integrity
- to monitor dozens of key performance indicators, individually (rank-ordered by criticality), and in combinations, ranging from revenue leakage to tarmac delays to customer satisfaction
- to develop and implement metrics to measure the value of partnering in an alliance

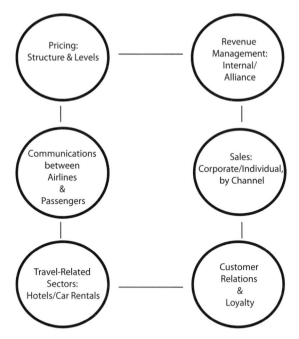

Figure 2.2   Technology Enabled Cross-Functional Integration within the Commercial Department

In addition to the above factors, the key value-adding insights that the advancing and converging technologies provide relate to (a) what competitors are doing in a particular market at a particular time through a particular channel, (b) what passengers are willing to pay for different components of the unbundled service, (c) how the brand is being impacted, positively or negatively, through the social networks, and (d) how an airline can communicate interactively with *each and every passenger* in a timely manner and with relevant information and options. Moreover, key performance indicators can be made available through a Web-based environment. The next two chapters provide numerous examples of how forward-thinking airlines (and a few airports) are using (Chapter 3) and could use (Chapter 4) new-generation technologies to change their business models to adapt to the intensely turbulent and ambiguous environment.

A smart combination of the aforementioned passenger- and operations-oriented technologies might even generate new business models. From the 1970s through to the 1990s the "East

Coast Shuttle" proved to be a highly successful business model, as it was convenient for the passengers (no bookings required) and profitable for the airlines, since it involved only as much capacity flown as required. Imagine a "Web 3.0 Shuttle" where passengers can check-in to the next possible flight anytime between 120 and 60 minutes before departure time. There could be only two fares: "Must Go" and "Can Be Moved To Next Flight." Airlines then could fill up planes with the second category up to 100 percent load factor and only pull a backup airplane when the backlog provides enough revenue to cover the variable cost of the backup flight. Reservation and ticketing systems, yield management, and other complex information and technology systems could be replaced by the combination of a mobile phone check-in system and an ad-hoc cost calculation program.

Takeaways

Key examples of consumer technologies include customized search engines, mobile applications and smartphone capabilities, social media and user-generated content, intelligent virtual assistants, and context-aware and location-based mobile applications. These represent the building blocks for acquiring information on customer behavior and, in turn, enabling the implementation of information-based business strategies. In addition, technologies are advancing to improve an airline's internal planning activities such as (1) developing an integrated and holistic schedule from the view point of the airline and its alliance partners, (2) diving deep into the core causes and multidimensional effects of airline disruptions, and (3) monitoring dozens of key performance indicators individually (rank-ordered by criticality) and in combination. Passengers and airline IT managers will most likely move from traditional PC- and server-based applications to hosted Web applications made possible by cloud computing with benefits including connectivity, scalability, dynamic provisioning, and per-use pricing.

Notes

1 "Server Virtualization: Solving Real Air Transport Issues," *New Frontiers Paper*, SITA, 2010.

2 Peter Yesawich, Ypartnership, Presentation at the OSU International Airline Conference, Seeheim, Germany, May 2010.

3 Michael Tasner, *Marketing in the Moment: The Practical Guide to Using Web 3.0 Marketing to Reach Your Customers First* (Upper Saddle River, NJ: Pearson Education, 2010), p. 11, also pp. 1–13.

4 Vanessa Fox, *Marketing in the Age of Google* (Hoboken, NJ: John Wiley, 2010), pp. 10–11.

5 Based upon an example given by Rick Mathieson in his book: Rick Mathieson, *The On-Demand Brand: 10 Rules for Digital Marketing Success in an Anytime, Everywhere World* (NY: AMACOM, 2010), p. 233.

6 Rick Mathieson, *The On-Demand Brand: 10 Rules for Digital Marketing Success in an Anytime, Everywhere World* (NY: AMACOM, 2010), p. 235.

7 www.hipmunk.com/

8 www.airfarewatchdog.com/

9 www.insidetrip.com/

10 Rob Torres, Google, Presentation at the OSU International Airline Conference, Seeheim, Germany, May 2010.

11 Roman Friedrich, Michael Peterson, and Alex Koster, "The Rise of Generation C," *strategy + business*, Booz & Company, Issue 62, Spring 2011, p. 57.

12 Clayton Reid, MMG Worldwide, Presentation at the OSU International Airline Conference, Seeheim, Germany, May 2010.

13 www.apple.com/

14 Rich Jaroslovsky, "The iPad Isn't Just Fun and Games," *Bloomberg BusinessWeek*, April 12, 2010, pp. 18–20.

15 Stephen Williams, "New York Auto Show: Hyundai's iPad Connection," *The New York Times* (online edition), April 1, 2010.

16 Jeff Bennett, "Mercedes to Give iPads to Dealers," *The Wall Street Journal* (online edition), May 25, 2010.

17 "Malaysia Airlines invests RM450 million to strengthen benefits to customers," Source: Malaysia Airlines, *Air Transport News*, 21/06/2010.

18 "11 innovative airlines to keep an eye on in 2011," www.airlinetrends.com, 25 January 2011.

19 http://www.google.com/support/mobile/bin/answer.py?hl=en&answer=166331/

20 Rick Mathieson, *The On-Demand Brand: 10 Rules for Digital Marketing Success in an Anytime, Everywhere World* (NY: AMACOM, 2010), p. 192.

21   Mitch Joel, *Six Pixels of Separation: Everyone is Connected. Connect Your Business to Everyone.* (NY: Business Plus, 2009), pp. 238 and 241.

22   http://itunes.apple.com/us/app/price-check-by-amazon/id398434750?mt=8#

23   Dan Butcher, "Sephora unveils multiplatform mobile shopping experience," *Mobile Commerce Daily*, August 4, 2010 and Giselle Tsirulnik, "Sephora redefines shopping for loyalists with unique mobile app," *Mobile Commerce Daily*, October 8, 2010.

24   Kevin J. O'Brien, "Mobile TV's last frontier: U.S. and Europe," *The New York Times* (global edition), May 31, 2010, p. 13.

25   Based upon material by Cindy Krum, *Mobile Marketing: Finding Your Customers No Matter Where They Are* (Indianapolis, IN: Que Publishing, 2010), pp. 133–152.

26   Rick Mathieson, *The On-Demand Brand: 10 Rules for Digital Marketing Success in an Anytime, Everywhere World* (NY: AMACOM, 2010), p. 192.

27   Jena McGregor, "USAA's Battle Plan," *Bloomberg BusinessWeek*, March 1, 2010, pp. 40–3.

28   John Arnold, "Mobile App or Mobile Website?" www.entrepreneur.com, April 5, 2010.

29   Erin Biba, "Inside the GPS Revolution: 10 Applications That Make the Most of Location," *Wired Magazine*, January 19, 2009.

30   Diane Brady, "Social Media's New Mantra: Location, Location, Location," *Bloomberg BusinessWeek*, May 10–16, 2010, pp. 34–6 and Brad Stone and Barrett Sheridan, "The Retailer's Clever Little Helper," *Bloomberg BusinessWeek*, August 30–September 5, 2010, pp. 31–2.

31   David A. Kaplan, "Apps: Hot Course on Campus," *Fortune*, May 24, 2010, p. 22.

32   "Where are you?" *The Economist*, August 28, 2010, pp. 53–4.

33   According to the United Nations Cyberschoolbus Document, Mashable.com, and Apple, Inc. as reported in *Airways*, February 2010, p. 53.

34   Statistics reported in this section are derived from the following source: "A world of connections," A special report on social networking in *The Economist*, January 30, 2010, pp. 3–4.

35   Statistics reported in this section are derived from the following source: "A world of connections," A special report on social networking in *The Economist*, January 30, 2010, pp. 3–4.

36   Chris Boudreaux, "How to Develop a Social Media Strategy," in *The Social Media Management Handbook: Everything You Need to Know to Get Social Media Working in Your Business* edited by Nick Smith and Robert Wollan with Catherine Zhou (Hoboken, NJ: John Wiley, 2011), pp. 16–35.

37 Kevin Quiring, "Social Media ROI: New Metrics for Customer Health," in *The Social Media Management Handbook: Everything You Need to Know to Get Social Media Working in Your Business* edited by Nick Smith and Robert Wollan with Catherine Zhou (Hoboken, NJ: John Wiley, 2011), pp. 36–53.

38 Anatoly Roytman and Joseph Hughes, "Creating and Implementing a Social Media Technology Platform," in *The Social Media Management Handbook: Everything You Need to Know to Get Social Media Working in Your Business* edited by Nick Smith and Robert Wollan with Catherine Zhou (Hoboken, NJ: John Wiley, 2011), pp. 189–208.

39 Brian Bergstein, "Inventing new technologies and markets," *technology review*, published by MIT, March/April 2011, p. 39.

40 www.groupon.com/

41 Rick Mathieson, *The On-Demand Brand: 10 Rules for Digital Marketing Success in an Anytime, Everywhere World* (NY: AMACOM, 2010), pp. 21–30.

42 Rick Mathieson, *The On-Demand Brand: 10 Rules for Digital Marketing Success in an Anytime, Everywhere World* (NY: AMACOM, 2010), pp. 74–5.

43 www.flightcaster.com/

44 www.gateguruapp.com/

45 Nick Bilton, "Helping Travelers Find Internet-Connected Flights," *The New York Times* (online edition), October 15, 2010 and Alexia Tsotsis, "HasWifi Shows You Which Flights Have Wifi," www.techcrunch.com, October 13, 2010.

46 www.seatguru.com/

47 Kristina Brambila, "The Future of Computing," *PC Today*, May 2011, Vol. 9 Iss. 5, p. 24.

48 "Hospitality 2015: Game changers or spectators?" A report by Deloitte LLP, 2010, p. 33.

49 "Hospitality 2015: Game changers or spectators?" A report by Deloitte LLP, 2010, p. 33.

Chapter 3
Current Information- and Technology-Enabled Initiatives

The air travel process for a passenger consists of numerous touch points, as illustrated in Figure 3.1. Points of integration within this process, enabled by technology, can help airlines find solutions to problems (a) expressed by passengers and (b) experienced by airlines themselves. The first chapter reviewed a few historical points of integration (such as interline agreements, hub-and-spoke systems, passenger O and D-based revenue management systems, and strategic alliances). In the process of identifying and implementing such points of integration, airlines also made some money, at least, in the short term. However, as most airlines began to deploy the same points of integration (such as developing hubs and joining alliances), innovations became commoditized and the incremental profit began to disappear. Simultaneously, new problems (faced, again, by passengers and airlines) began to appear. On their part, passengers started to become more demanding as their lives became more complex and time constrained, as they became more empowered with information about products and prices provided by the new wave of technology, and as their expectations began to be raised by the positive experiential aspects of the services provided by trend-setting companies such as Apple. As for airlines, their problems also became more severe as competition became more disruptive, not only among the traditional full service carriers, but also as the distinction began to blur between the low cost carriers and the full service carriers.

Figure 3.1   Holistic View of the Traveler

The emerging duality of the problem, faced by businesses and their customers, is similar in other industrial sectors. However, some of these sectors were early adapters of the new-generation technology (the mobile Internet, advanced search engines, and social media) to find solutions to the duality of their problem. To their credit, airlines did try to follow a similar path but their initiatives were limited for, at least, four basic reasons:

- First, problems experienced by passengers are immensely complex and massively varied. Some airlines have simply not been able to grasp the serious consequences of passenger-experienced problems, let alone provide acceptable solutions. This problem seems to occur less with passengers of low cost carriers as these airlines neither promise a lot nor have complex processes with many opportunities to fail. Here is a repetition of the problem quoted a number of times in this book (and as illustrated in Figure 3.2). The first short-haul segment of a long-haul intercontinental flight is delayed resulting in the passenger missing the long-haul connecting flight. The airline simply tells the passenger that the problem has been solved as he has been booked on the same flight the next day. On the side of the passenger, the consequences could range from completely missing an important business meeting, to being unable to participate in a wedding, to missing the connection with a cruise ship. Being booked on a flight 24 hours later may simply not be an acceptable solution from the viewpoint of the passenger. The value that a passenger assigns to the solution of a particular problem will undoubtedly be different than the value assigned by an airline.

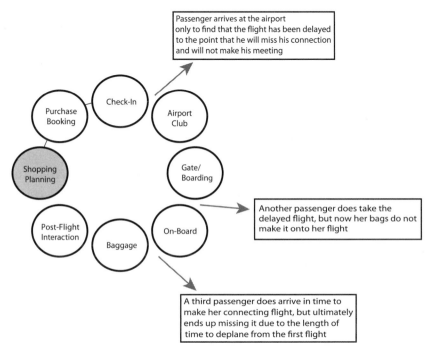

Figure 3.2   View of the Traveler: When Disruptions Arise

- Second, traditional IT departments at many airlines are focused on (a) simply maintaining their legacy systems and applications, (b) trying to integrate the legacy systems with the new-generation technologies, (c) trying to reduce not only the airline's operating costs, but also their own departmental budgets, and finally, (d) finding ways to improve the service provided to passengers. Again, to their credit, while working within numerous constraints (ongoing challenges relating to the legacy systems, low IT budgets (in relation to the demand to develop an airline-specific IT system), lack of full support from the C-suite, let alone from the Boards), traditional IT departments have achieved some successes, illustrated by the examples given in this chapter. The problem is twofold. At some airlines the improvement in passenger service has a lower priority than the need to maintain the legacy systems and applications. The successes that have been achieved by traditional IT departments have been achieved in a piecemeal manner as opposed to a holistic approach.

• Third, most airlines continue to gather and analyze data within the framework of their conventional rules or requirements established by governments. Data on delays is kept on arrivals and departures within a given number of minutes of scheduled times, say 15 minutes. Even when a plane does arrive at a gate within this time limit, some passengers miss their connections if there is no agent to connect the jet way to the aircraft for 10 more minutes. Then there are times when the plane arrives on time, and the jet way is connected on time, but passengers seated at the back of the single-aisle plane cannot deplane in time to make their connections, as depicted in Figure 3.2. Some passengers in the front, who are not in a hurry, take too long in gathering their baggage from the overhead compartments. The point is that data gathered and reported is more airline-operation-centric, and less passenger-centric. Even when an airline does keep track of passengers missing connections at a hub, is the data divided by the cause of the problem? It is ironic that technology is available to keep track of such data in microscopic detail and in real time. A passenger would reduce her stress level—information-based mass customization—while waiting to deplane one plane to connect with a second plane by knowing not just if the connecting flight is on time, but the current gate location, the shortest time, and direction to the gate. Technology is even available to include a prediction capability, such as the likelihood that the connecting flight will have a gate hold due to congestion on the ramps, taxiways, or runways. Most passengers carry mobile phones and this communication would improve the customer experience. However, while some forms of advanced information about disruptions could be provided to passengers, it may be counterproductive to do so depending on the circumstances. One potential concern of an airline could be the possibility of an air rage resulting from the notification of a problem to passengers during the flight.

- Fourth, passengers are placing a premium on the experience aspect of the entire trip, not just the part of the trip under the control of an airline, even if one assumes that an airline has been able to integrate its services at each and every touch point within its own part of the travel chain. As with airlines, airports have also been trying to use emerging technology to improve the part of the services they provide to their customers (airlines) and their customers' customers (passengers). Examples of technologies being used by airports include Flight Information Display Systems and Wi-Fi. Airports are now experimenting with the use of such technologies as Radio-frequency Identification and biometrics. However, as with airlines, there are limitations as to how far airports can go in bringing about an improvement in the totality of the trip. Not only are there gaps between the systems and applications provided by airlines and airports, there are also many services provided by other organizations such as by governments relating to security and immigration control.

Notwithstanding the aforementioned constraints faced by airlines and, to some extent, airports, both sectors must be credited with numerous technology-led achievements described in the following sections. On the other hand, the successes have not only been piecemeal but they have been valued according to value systems assigned by service providers, not passengers. Moreover, even when service providers substantiate their value systems based on the results of surveys, conclusions are drawn from averages. Once again, technology is now available to receive information on micro-segments, by flights, by cabin class, by point of origin or destination, by date, by crew member, and so forth. It is the deployment of such technology that will enable an airline to move from efficient transactional to effective interactional, and holistic, initiatives, and thus the transition from product centric to customer centric. While still focused on product centricity, airlines have begun to move toward customer centricity such as through the personalization of experience. For example, United Airlines offers a limited number of options for its passengers to customize their travel experience, as depicted in Figure 3.3. The limitations are based on costs and infrastructural constraints. While some of these options offered are not new (travel insurance), others are (priority boarding that enables passengers early access to overhead bin space). Moreover, passengers can choose and reserve meals prior to their flight, and passengers can choose to pay for and receive additional miles for their flight. Passengers can choose individual product attributes, or in an attractive bundle of benefits featuring:[1]

- bonus mileage
- up to two free checked bags
- priority check-in line access
- priority security line access
- priority boarding
- additional legroom

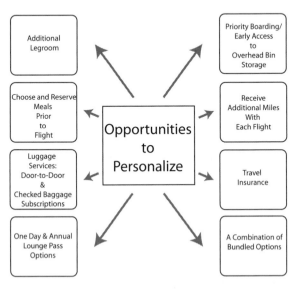

Figure 3.3   **Progress Toward Customization (based on the information available through united.com/traveloptions)**

Reducing Costs and Improving Passenger Experience

Evolving technology has been enabling airlines to reduce their operating costs and improve the service provided to passengers in many areas, ranging from self-service check-in devices and services, to questions being answered by virtual assistants, to door-to-door transportation of baggage.

Check-in and Boarding

Check-in is one area where airlines and airports have clearly deployed advancing technology to reduce costs and improve passenger service. Think about the old method of check-in in which passengers conducted their transactions at airport counters. Modern technology is enabling passengers to check-in via other channels that are more convenient. Typically, passengers can now check-in via the Web (by entering a booking confirmation number or frequent flyer number), make seat selections, and receive boarding passes. Boarding passes can be printed off-site or sent to passengers' mobile devices, where the device itself can even be used as a boarding pass in some cases. While boarding passes can be sent to a mobile phone, they could also be sent to a wireless printer at any location. Technology is now available that can enable an airline to send a boarding pass through email that could go directly to a printer located in the business office of a passenger's hotel. Frequent flyers can perform a one-time registration for SMS check-in and can then receive an SMS 24 hours prior to flight departure. A passenger can even make a seat selection easily by typing letters such as "W" for window and "A" for aisle. Some airline agents are even going mobile. They are using handheld devices to issue baggage tags and boarding passes in an attempt to manage wait times. KLM has even implemented a program which allows its passengers to self-tag and check-in their baggage.[2] Alaska Airlines has taken the self-service concept one step further: the carrier has implemented a program at a few airports in which the traditional ticket counters are replaced with check-in kiosks and luggage check stations.[3] Airlines are now beginning to offer mobile check-in and paperless boarding for international travel.

On the part of airports, managements have begun to use technology that can improve passenger service relating to the use of self-service for check-in. Next, the use of "Flight Transfer Kiosks" has been increasing at a rapid rate. They are now not only working closely with airlines in the area of self-service check-in devices, but they are also exploring ways to expand the use of self-service in other areas including tagging baggage, security clearance, and border control. The key lies in learning about passenger travel behavior. While it is true that self-service check-in facilities have improved passenger satisfaction in one area and reduced costs for airlines (and to some extent even for airports), the new systems have also made it more difficult for airlines and airports to know when passengers would be physically arriving at the airport, since passengers would already have checked in at off airport-sites.[4] Lack of such information will have an impact on the traffic flow (passengers and baggage) at various locations, creating a need for collaboration among different aviation entities at airports. Again, technology is available to proactively track and trace passengers in real time to improve productivity of airline and airport resources as well as security. One example is Bluetooth technology. While Bluetooth technology itself is not new, it is now being used in the airline industry, such as through initiatives led by SITA, to track movements to better understand the behavior of travelers, and ultimately, manage the time spent waiting in line throughout the airport. A recent survey conducted by SITA indicated that travelers value short lines as well as punctuality as key components of their travel experience.

Today the standard information carrier for e-boarding passes is the 2-D bar code. It is an improvement over the magnetic strip, as it can be produced on every printer and mobile phone screen; however, it needs an optical connection to a reader and can only be read one at a time.

In Asia, advanced subway passenger handling systems use RFID technology that enable the processing of much larger passenger volumes. Also, as cargo warehouses show, RFID tags can be easily located within a building, allowing a better tracing of passengers as they move towards the departure gates. And finally, RFID chips integrated in mobile phones could work as boarding passes, rental car keys, and hotel room keys, as discussed in Chapter 2. One carrier that is already pursuing this technology is Air New Zealand. The carrier offers its frequent flier passengers the opportunity to attach a RFID sticker to their mobile device that may be used to check in on select flights (domestic and short-haul), check into airport lounges, and even self-board at the gate.[5] Qantas has introduced what they refer to as their "Next Generation Check-in" that includes a chip in both the frequent flier card and the baggage tag of a passenger. This initiative is in response to passengers' concern regarding the time spent in check-in lines.

As SITA recognizes, thus far, mobile self-service in the air travel sector has mostly involved the check-in part of the travel experience. However, as SITA points out, mobile self-service could be rolled out more fully to encompass more touch points and at a greater level, including booking reservations, managing reservations, filling out forms (such as landing cards and visa applications), and managing delayed or lost baggage.[6] For more on the future of mobile in terms of opportunities for both airlines and airports, see Chapter 4.

Virtual Assistants

Technology is enabling airlines to delve into the deployment of virtual assistants to improve the user experience of passengers while on an airline's website. In the US, Alaska Airlines and Continental Airlines are utilizing such technology to enhance the customer experience on their websites by offering virtual assistants to answer questions and direct travelers to the information that they are seeking on the carriers' websites. In the case of airlines, their virtual travel guides answer general questions, allowing their customer service call center staff to focus upon solving more complex issues. Travelers are able to type questions in a window on the airline's website to which the virtual assistants respond verbally, asking follow-up questions when appropriate. The virtual assistants also give written responses as well as display the most relevant page on the website with respect to the topic of the question. Furthermore, these assistants relate to the human nature of trust; they actually have some depth to them and therefore customers are able to actually ask them some personal questions which demonstrate this quality. More importantly, they are able to understand intent, which is *the* critical success factor to virtual assistants. The self-service technology provides travelers with a personalized experience that provides relevant and timely information.[7] Think of the next stage of development in this area. Technology could enable questions to be even asked in different languages and virtual assistants in different sectors of the travel industry could collaborate (within the airline, hospitality, car rental, and tourism information centers) to improve the totality of the travel experience.

A different example of a virtual assistant is Siri, a mobile application that was built because the developers believed that there was an easier way of getting things done. It can help a user find and plan, whether it be finding a special venue for dinner or obtaining movie tickets. A user can even arrange for a cab through Siri. A user may type or speak her requests. While Siri is still in its early stages of development, the idea is for the application to eventually manage many of the personal details of a user's life. Currently the application works on iPhone and iPod touch devices in the US, but plans are underway for additional mobile platforms.[8] Clearly, such a service could add significant value for passengers to improve their airport experience. Devices could display detailed maps of airports with interactive and analytic functionalities. A passenger could ask not only for directions to a certain gate from his existing location but also the amount of time it would take for him to get to the gate.

Even though technology is available and passengers have had positive experiences with the use of virtual assistants, only a few airlines have incorporated them into their websites. On the other hand, whereas airlines have at least begun to upgrade their websites in this area, now airports can explore the deployment of virtual assistants.

Collaboration and Convergence

There are a myriad of touch points involved in the travel experience as shown in Figure 3.1. While airlines have become pretty good at improving the passenger experience at these steps individually, the challenge lies in tying all of these parts together. However, the activity of tying the parts together is, in fact, where the opportunity lies (the new point of integration for adding value and making money), and technology can help airlines capitalize on this point of integration opportunity. If an airline fails to get a passenger's luggage on-board the flight (as illustrated in Figure 3.2), the airline already has that information, and could use technology to alert the passenger well before the passenger is standing at the baggage claim waiting for luggage that is not even there. Some airlines have now begun to develop such a vision of an integrated approach, one that involves not only changes to internal processes within the organization, but also collaborating with airports that are working at the same problem from a different direction. Delta and OTG Management, an airport restaurant operator, have collaborated and leveraged technology to augment the travel experience by creating a new type of dining option at New York's JFK Airport. Dining areas have been created near the gates that are equipped with Apple iPads featuring a custom application that enables Delta passengers to order food and beverages. The food is brought right to the traveler, so that they do not have to leave the gate area.[9]

Some airport managements have not only been exploring the use of RFID, but one (Amsterdam's Schiphol) has even begun to experiment with the deployment of baggage loading robots. The bottom line is that (a) airlines (and airports) can make profits at the point of integration, and (b) technology can help them achieve this integration. One challenge to consider in the integration of services is the large number of systems and providers involved in a typical passenger's trip. If a passenger travels often on the same airline, closer integration may be possible. However, if a passenger travels on many different airlines on a regular basis, it is unlikely that a passenger would be receptive to sophisticated systems that require them to use dedicated applications or provide a lot of personal data. Consequently, there may be an opportunity for external intermediaries who may have a clearer role in providing these services for at least one major sector of the traveling community. Another consideration is the challenge that such sophisticated systems place on legacy applications, and thus the complexities in implementing them.

The above examples, relating to airlines and airports directly, are just scraping the surface of improving the total travel experience through convergence and collaboration; much more can be done. A service that has leveraged technology to augment the passenger experience is TripIt. The service collects, collates, updates, and distributes information, digitizing all of a passenger's itineraries. The service was launched in an effort to solve travel problems, as once the information is all in one place it can be used to provide additional information such as maps and weather, as well as share the traveler's information with friends and family, or even co-workers, all at the discretion of the passenger. Furthermore, the service can push this information out to mobile, social networks, calendars, company groups, as well as store it all in one place available to print, again, at the discretion of the traveler. Furthermore, such initiatives can allow travel companies to access rich and relevant information on a traveler. If a company can secure the traveler's trust, it can ascertain such information as purchase history (and therefore preferences), current location, etcetera.[10]

Take the case of Frommer's Unlimited, a service company that has leveraged technology to augment the passenger experience. The company has been known for years for its travel guide books. However, the company now also offers a service that enables travel service providers such as airlines and hotels to augment their value propositions. The service leverages Frommer's travel expertise to help augment a travel provider's offerings. Following are just two examples. British Airways collaborated with Frommer's Unlimited in an effort to improve the carrier's website. Frommer's Unlimited created travel guides that were integrated into British Airways' website, written in the tone of the carrier and in alignment with the brand, and provided travelers with relevant content such as information on events occurring in the destination city, restaurant reservation bookings, and even weather forecasts. Furthermore, Frommer's Unlimited leveraged technology to match the airline's promotions with events occurring in corresponding cities. In one online promotion for US audiences, they featured an interactive map of London to not only illustrate area attractions, but also to highlight hotels participating in the airline's promotion. Hilton Hotels desired to improve their website's content in an effort to better market their properties and the corresponding locations. Frommer's Unlimited created descriptions of the properties while adhering to the brand's guidelines. In addition, it created corresponding destination guides which contained content that was tailored to the hotel's customer demographics. Not only did the guides present relevant information on topics such as sightseeing and nightlife, but they also were integrated with airport guides, events calendars, and weather forecasts.[11]

Consider now an example of an integrated service, enabled by convergence: a passenger's flight is delayed, an alert about the delay gets pushed to the traveler's mobile device, and *only relevant* alternatives available are presented to the traveler automatically. Furthermore, the traveler's rental car pick-up is rescheduled, as well as the traveler's business meetings, and even the traveler's hotel is notified. Companies are already trying to capitalize on the concept of convergence with personalized, one stop shopping of services. Rearden Commerce offers, for example, a productivity tool which literally is an online personal assistant. The purpose of the tool is to pull everything that a traveler may need together into one single system. The tool eliminates (1) the need for a consumer to go to numerous websites to book various elements of an itinerary (such as airline, hotel, rental car, restaurant reservations, events), and (2) the need to remember the username and password for each site. Furthermore, the system "learns" a consumer's preferences and obtains the relevant services she needs, including travel planning, dining reservations, and even Web and audio conferencing, as well as a traveler's company travel policies, restaurant reviews, and maps and directions. Finally, the system has the capability to help a user manage her busy life through functionalities including email invites, calendar updates, and alerts sent directly to her mobile. Such a service truly capitalizes on personalization-enabling technology to provide the traveler with a "virtual" assistant.[12] Much like the online personal assistant developed by Rearden Commerce, the future of airlines could involve the development of the travel concierge, a possible next point of integration.[13] Such a development could improve customer experience, reduce costs, and develop new streams of revenues. As suggested by some analysts, airlines might even offer three levels of concierge services during a passenger's trip and charge a fee accordingly:

- Premium: Person-to-person, real-time contact, using a special key on a mobile device that can connect a passenger to a concierge who has access to all relevant information concerning the passenger's travel status. The concierge can reroute and rebook in the event of disruptions. The concierge can also provide information regarding travel-related services and their payments.
- Standard: Personalized service, using email and voicemail to collect requests from travelers asking for the same services and assistance as in the premium case. One concierge can support many passengers at the same time and thus the service can be produced at a lower cost. However, passengers should be willing to accept a delay in response to their requests that will be answered via their mobile device.
- Economy: Self-service website that provides various levels of information and functionality.

Emerging Signs of Door-to-Door Travel

Although door-to-door travel (as opposed to airport-to-airport), similar to the services provided by courier service companies, is far from becoming a reality, the enabling technology is here. Some travel providers have already begun to explore the concept of door-to-door service, although it is still in its early stages. Southwest Airlines joined the Resort Airline Check-In service at Walt Disney World Resort, in which guests have the ability to check their bags all the way through to their final destination before even departing from their hotel. Virgin Atlantic is another travel provider that is going beyond gate-to-gate service. Virgin has check-in facilities at Downtown Disney in Orlando which allows passengers to check-in for their flights and obtain their boarding passes, as well as drop off their luggage to give them more time to enjoy the last day of their vacation. Virgin Atlantic also offers similar programs in Las Vegas, with off-airport check-in kiosks at locations such as the Luxor Hotel, the Venetian Hotel, and the Las Vegas Convention Center. Again, the idea is to free the passenger of having the burden of dragging bags around as well as to give the passenger an opportunity to maximize the last moments of her vacation time.[14] Swissair implemented such initiatives in the 1990s within the whole country. Although passengers are extremely in favor of such initiatives the challenges have now become security related.

United Airlines has taken the concept of door-to-door baggage one step further with its program which is actually called "Door-to-Door Baggage" (as depicted in Figure 3.3). This program enables a traveler to avoid the hassle of hauling awkward items such as bikes, golf clubs, or skis through the airport. Rather, the traveler schedules his shipment online, and then has the choice of either dropping the luggage off at a shipping center or having FedEx pick up the shipment, that is then delivered directly to the traveler's destination that could be a business, hotel, or residence. The traveler pays a set rate each way for each item.[15] Similarly, All Nippon Airways has implemented a service, for a fee, that allows passengers traveling domestically to have their baggage delivered/picked up to/from their home or office.[16]

While some airlines are focusing on the baggage component of the travel experience, other carriers are focusing on linking ground transportation with air travel. Etihad Airways offers a service called "Etihad Chauffeur," which is available to its first-class and business-class passengers. The program, that includes luxury vehicles, personal drivers, and even water and newspapers for passengers, is currently available in a dozen cities worldwide. Although the door-to-door part of the trip for premium travelers is not a new concept, Etihad is striving to offer passengers the same level of service in its ground service transportation as it does on-board its aircraft (extending the brand).[17] In the case of Virgin Atlantic, on top of its baggage service previously described, it raises the bar for the concept of door-to-door travel service with its Upper Class program, that includes complimentary chauffeur-driven car service to the airport and at Heathrow, and a special drive-through check-in program at the airport that enables the passenger to go straight to the first-class lounge, referred to as the "Clubhouse." Complimentary limo service is also provided at the destination.[18] All Nippon Airways features helicopter or limousine service to Central Tokyo to first-class passengers who travel on select flights.[19] All of these different travel providers within the value chain have come together to collaborate on end-to-end travel processes rather than just concern themselves with their own part in the chain. While this is a step in the right direction, airlines can take the next step, that is, leverage technology to be able to provide such solutions to the next level of travelers, not just the top-tier, first-class, or even business-class travelers.

Beyond limousine and baggage transfer there are many more opportunities to better interlink air and ground transportation. Some airports like Hong Kong have specific rail feeders that also provide downtown check-in facilities. Airlines could offer many more locations for passengers to check in, such as off-airport rental car centers, long-term parking lots, or valet drop-off services right at the curb of a terminal. Perhaps, a company such as Foursquare could provide the "mating" technology.

On the airport side, new-generation technologies along with the rise in smartphone usage make the concept of an "intelligent airport" conceivable in the near future. One key component of an "intelligent airport" is a true knowledge of its ultimate customers (passengers) and their pertinent information, allowing the facility to predict peak demands, minimize lines, plan staff levels as well as facility amenities such as parking accordingly, and most importantly, help travelers navigate from ground transportation to air transportation in real time. Three areas that are currently holding this concept back from becoming a reality are (1) inefficiency, (2) ineffectiveness, and (3) lack of agility. Like airlines, airports can (1) integrate their fragmented IT systems, (2) implement real-time decision-making tools, and (3) implement predictive analytics.[20]

Embracing Social Media

Social network sites and mobile devices are becoming important elements in the channel mix. What is truly key to the airline industry is the relationship between these two elements. Not only can passengers go home and broadcast their travel experiences, whether it be positive or negative, but now travelers have the ability to broadcast to their networks right from their mobile device while en route. This development can have a global impact on airline brands. Consequently, airlines must develop effective strategies to not only sell more through these new channels but also manage their relationships with customers in real time.[21]

While most airlines have recognized the benefits that can be derived from social media, some are utilizing the channel more comprehensively than others. American Airlines is now using the channel to communicate with the carrier's loyalty program members. Moreover, the airline views the channel as a way that members can interact with each other as well, such as through sharing tips with fellow members on how to earn more miles. JetBlue is reported to have over a million followers on Twitter. This kind of following requires significant airline resources, such as a dedicated team within the corporate communications department of the airline that monitors the feedback from all passengers, every day. The airline dedicates such resources to social media, presumably, because it sees an opportunity in the areas of customer satisfaction and brand loyalty. The carrier presumably views Twitter as a means of carrying out its belief of being a customer service company above being an airline. The social media tool enables the carrier to monitor customer service performance, both positive and negative, in real time. While the airline may not be able to point to a direct impact on the bottom line from social media, it does recognize that the areas of customer satisfaction and brand loyalty certainly do have an impact on profits. Virgin America, like JetBlue, recognizes the linkage between social media and branding. Virgin noted that social media sites were driving more traffic to the corporate website in Europe. Alaska Airlines is another carrier that sees social media driving traffic to its website: the airline observed a significant increase in traffic to its website after the carrier ran a "mystery fare" sale via Twitter. More importantly, the airline learned from this experience that there is a fundamental shift in the manner in which consumers react to advertising messages. As one analyst has noted, consumers no longer patronize a brand because the brand tells them to do so, but rather because their friends tell them to do so.[22]

The social media phenomenon is also being adopted by all kinds of airlines worldwide. AirAsia is becoming a leader in terms of engaging with passengers through social media. Each time that the carrier launches a new destination, AirAsia creates a special site for it so that travelers may learn about it. AirAsia X (a subsidiary of AirAsia), recognizes the power of social media in terms of building brand ambassadors. An executive at the airline even goes so far as to say that social media may even be more effective in building loyalty than traditional airline frequent flier programs. Another airline, WestJet has also incorporated social media as part of its overall communications strategy. When the failed terrorist attack on a Delta aircraft occurred on Christmas Day in 2009, the US Transportation Security Administration ordered all US-bound carriers to implement secondary security measures. In response to these new measures, WestJet alerted their passengers via Twitter and Facebook. By doing so, the airline was able to not only quickly inform its followers, but also respond to their questions and concerns in real time.[23] In Europe, both airlines and airports leveraged social media to help manage the 2010 volcanic ash cloud crisis.

While some airlines have not only embraced social media, they are actually using it to focus on certain niche markets. American Airlines launched a social network, BlackAtlas.com, as a result of focus groups with African-American customers. The site enables travelers to conduct fare and vacation searches, as well as obtain destination information, all customized to their interests. Similarly, AirTran has been targeting college-aged students by offering standby fares through airtranu.com. Finally, KLM offers virtual business communities, Club Africa and Club China, for passengers conducting business in those areas of the world.[24]

While social media has proven to be valuable for airlines in terms of engaging with passengers in real time, it is also a potential opportunity to sell more tickets while users are actually on their social networks. The social network arena may not replace current distribution channels, but it can serve as yet another vehicle to capture additional traffic, especially if it is a component of a carrier's overall holistic strategy. AirAsia, a carrier that recognizes the value of real-time feedback that social media enables and its importance to building the brand as previously discussed, also uses the channel for exclusive sales campaigns for its social network followers. Recently, the carrier ran a one-day promotion that featured a two seats for the price of one special for its Facebook fans.[25]

Consider another aspect of social media, online videos, that has been used successfully by other businesses to engage consumers in new ways. One of the most popular video sites is YouTube. YouTube was acquired by Google in 2006. Google is reported to have 43 percent of the online video market, while YouTube is reported to make up 99 percent of that.[26] One example is Procter & Gamble, which developed an online video for its Old Spice brand that is reported to be one of the ten most-watched online video advertisements on YouTube in 2010. While YouTube is well known for all of its famous viral videos, the medium can be used by businesses as a means of product demonstration, employment videos, customer testimonials, and so on. Some analysts recognize that video can be a more effective means of engaging consumers rather than just written material. It can be an effective means of educating and informing consumers.[27] Two critical success factors reported are first, to keep the videos short, and, second, to make it intriguing and captivating.[28] The point is to create something that will not only engage consumers, but will make them want to share it with others. As one author notes, social media success should not be considered by the number of consumers who received the message, but rather by the number that found it to be intriguing, and more so, found it remarkable enough to share it with their friends.[29] Some airlines have already pursued this channel for safety videos. An example outside of the airline industry is Freescale, a spin-off of Motorola. Freescale has implemented YouTube videos as a part of its recruiting strategy; engineers at the company create videos for prospective employees to give them more insight into the company.[30] There is also IKEA that has used YouTube to offer "How To" videos for its products as well as its home planner tool.

Within the airline industry, two carriers have particularly developed captivating videos that have resulted in a "wow" factor and generated a lot of buzz: Cebu Pacific and Air New Zealand. In the case of the low cost carrier, Cebu Pacific, the intent was to create a fun atmosphere through in-flight live entertainment. The flight attendants danced to current music while providing safety instructions. However, a passenger videotaped the act and put it on YouTube, where it became a viral success (ten million hits on YouTube and discussion within the global media).[31] Moreover, the carrier has now been enjoying the benefits of the viral success in terms of brand awareness and free marketing, highlighting the fun and friendly nature of the service. In the case of Air New Zealand, the intent is to illustrate the key product features, such as its new economy-class seats (see photo in Chapter 6) and in-flight entertainment through a little furry character named "Rico." While this marketing campaign has become somewhat controversial, it has also engaged consumers as they wish to follow him. Online videos are inexpensive to produce relative to television commercials. However, it is difficult to quantify their benefits in the near term. On the other hand, such social media has been reported to produce emotion, with the expectation to eventually grow revenue in the long term. The critical success factor is the necessity to incorporate online videos in the holistic social media strategy (as discussed in Chapter 2).[32]

Social media clearly provides amusement, which can lead to engagement and tap into the emotion factor. It also can be an efficient means of conducting day-to-day business such as customer service (such as through Facebook and Twitter). However, if the ultimate goal is to use this channel to grow business, there are certain requirements. First, there is a need to commit resources, people, and technology. Second, this commitment needs to be organization-wide, and consistent across the organization. Third, a carrier cannot expect an ROI in the short term.

Context-Aware and Location-Based Mobile Applications

Progressive airlines (as well as airports) have already begun to use technology to pursue context-aware and location-based mobile applications, to augment the customer experience with respect to connecting passengers. For instance, as one technology leader suggests, a text message could be sent to a passenger with a tight connection due to a late incoming flight informing her that a cart is waiting at her gate to take her to her next flight. At the same time, another connecting passenger who is also on the same flight but has a later connection could receive a pass to access the airport lounge via his mobile device.[33] For information on mis-connected passengers, see the example in the Appendix. SITA also recognizes the power of mobile agents in terms of augmenting the passenger experience. For instance, an agent could meet a passenger as she deplanes from her flight to inform her that her luggage will not be arriving on the same flight, then scan the passenger's baggage tags to obtain information on its status, and finally, use the mobile device to complete lost baggage forms if necessary. Furthermore, SITA recognizes that equipping agents with mobile devices could assist agents in terms of being able to access the same information as the passengers that they are serving.[34]

One travel technology provider is introducing products that have the ability to detect a traveler's location, a technique referred to as geo-localization that enables the delivery of targeted information and services. If a traveler is using her smartphone to book a flight, and she happens to be standing in London, then the system will assume that the passenger wishes to depart from London, and the screen will fill London as the departure city. Furthermore, if a traveler is trying to find a ticket office for Malaysia Airlines, the system will indicate offices in the London area.[35] Similarly, geo-localized tracking made possible by mobile technology could be used by airports in an effort to manage the flow of passengers, as they work to attain the "intelligent airport" as described earlier in this chapter.[36] Taking the concept of geographic location a step further would enable an airline to offer price-service options in real time. If a passenger has crossed the security line and is informed of a delay or cancellation for his flight, options could be provided for flights leaving from gates within that secure area as well as reservations being changed accordingly.

Virgin Blue is using mobile technology to help reposition itself from a low cost carrier to a carrier of choice for the business segment. In trying to attract this segment, the airline is working with BlackBerry and WorldMate (a travel itinerary service). While the more standard functions such as booking, checking-in, and managing flights will be available on most mobile devices, the BlackBerry version will feature additional advanced functionalities that other devices will not, not even the iPhone. One key feature of the application will be the ability to rebook automatically on alternate flights, eliminating the need to stand in line at the check-in counter in the case of flight delays or cancellations.[37]

Expedia is also using the mobile channel to target consumers with its iPhone application, TripAssist. This application gives travelers the ability to research destinations, flights, car rentals, and hotels. It also allows them to manage and update their itineraries, as well as look up alternate flight information and even view a seat map. Finally, the application provides free SMS and email updates. Features include a notice within 24 hours of travel; a screen pops up on the user's mobile indicating when and where he needs to be. Another feature is targeted to those who are picking up a traveler at an airport. The driver may receive either SMS or email updates for the traveler's flight. One click connects the user with Expedia customer service. The application was developed with the mobile limitations such as small screen size in mind in that it has a stripped-down and easy-to-navigate interface. The application was developed to simplify the complex travel process and meet the needs of travelers while en route.[38]

Emirates has been leveraging technology in the area of in-flight entertainment and connectivity. Through its information, communication, and entertainment platform, referred to as "ICE," passengers are currently able to send emails, send SMS communications, and make phone calls.[39] Emirates has also leveraged technology in an attempt to equip its cabin crew with passenger information in real time. The crew can access the carrier's CRM system through a digital device and therefore obtain information such as regarding the preferences of passengers.[40] On the other side of the world, Malaysia Airlines announced the iPhone application, MHdeals, demonstrating how airlines can exploit the technology commercially as a new distribution channel for ticket sales. The application uses GPS to determine a user's location and displays the nearest airports from which one can access great deals. A user simply holds the phone in front of her to view the surrounding airports and then chooses one to view flight offers. She can then book her flights through the MHmobile application.[41] Returning to in-flight entertainment, Air New Zealand is using its touch screen system for passengers to order food and beverages as well as listen to music and watch videos together with other passengers, while Virgin America is using its in-flight entertainment system, named "Red," for passengers to play games, shop, watch television, and even "talk" with other passengers through the seat-to-seat messaging feature. Furthermore, the IFE system features a music library in which passengers may even choose their own playlists. Red may also be used by passengers to order food and beverages; passengers input their orders via the IFE system and flight attendants in turn receive the orders via a tablet PC.[42]

Lufthansa has tried to help its passengers navigate through more of the travel experience than just the flight itself. The carrier has launched an iPhone application, Lufthansa Navigator which consists of a navigation system type of map, in an effort to assist travelers as they make their way through the Frankfurt Airport.[43] KLM is capitalizing on the location-based mobile application, Foursquare that was described in Chapter 2. Passengers checking in at Amsterdam Schiphol Airport via Foursquare randomly receive a "gift" based upon data which employees collect from passengers' social media accounts. Employees choose a personalized gift for the passenger, physically locate the passenger in the airport, and surprise the traveler with the gift. The information is then featured on KLM's Facebook page as well as its Twitter account.

As for airports, they can use mobile technology to improve resource productivity (by monitoring passenger flow), to raise customer satisfaction, and to generate additional revenue from concessions. Instantaneous, meaningful, and relevant information, enabled by technology, is leading to greater customer satisfaction and an increase in retail revenues by transmitting the relevant marketing message to the right customer at the right location and at the right time.

Where Do We Go From Here?

As this chapter highlighted, airlines and airports have been working hard to meet the ever changing needs of passengers in general terms, based on old-generation segmentation techniques and averages based on the results of passenger surveys. On the other hand, technology savvy passengers can almost "see" technology-enabled solutions to their *personal* problems. Ironically, game-changing technology is available right now to move to this next level of solving passengers' problems, *at a personalized level*. As described in the next chapter, one component of the step-changing technology can first be used to identify the multidimensional problems faced by different passengers and the same passengers at different times and at different locations. Another component can be used by an airline or an airport to integrate the initiatives within their own multifaceted divisions and functions. Another component relates to the forthcoming e-enabled aircraft with enormously increased on-board and external data communications, capacity and flexibility. In addition to the electronic flight bags and electronic technical logs capabilities (described in Chapter 1), e-enabled aircraft will also enhance the capability to provide improved passenger services right at their seats (shown, for example, in Appendix Figures A6 and A7).[44] Yet another component can be used to collaborate and integrate with other members in the travel chain to provide totally holistic solutions.

In closing, consider a few cases of how other companies have successfully recognized and met customers' needs by implementing innovations through new-generation technologies. There is no question that the airline business (for example, reservations) is far more complex than many other businesses (say the reservation process at restaurants, described below). Nevertheless, there are a few key insights.

Interactive Fitness Holdings (IFH)

- IFH has shifted the traditional gym product offering (working out on a traditional fitness bike) to a fitness-related service(by implementing technology to connect fitness equipment to virtual reality) after recognizing that many gym members become tired of their routine and lose interest quickly and therefore creating a differentiated service that meets the exerciser's needs including motivation and engagement. The company's line of Expresso bicycles are not just a product, but rather offer an experience such as visual scenarios via the Web in which users can engage on bike tours such as in the mountains or along the coast, bringing the outdoor workout experience into the fitness club. Moreover, it recognizes that the user experience is constantly evolving, and therefore the offering is automatically updated via the Internet in terms of courses and games available for the user. This not only keeps the consumer engaged, but it also is a positive attribute for the employees of a fitness club as they are not burdened with having to deal with implementing the upgrades. Furthermore, IFH has introduced software that enables users of their Expresso bikes to share milestones (accomplishing a particular bike ride, achieving a particular mileage level) via their social networks. It even has a feature that allows friends of users to "challenge" each others' achievements. As the company's website highlights, it is working to bring fitness into the twenty-first century, and appears to be successfully doing so by applying interactive and virtual technology to traditional cardiovascular exercise.[45]

OpenTable

- OpenTable has shifted the traditional labor-intensive restaurant process of managing reservations, coordinating tables, and managing customer relations to a real-time online service solution after recognizing that not only could technology be implemented to streamline these processes, thus enabling restaurants to better manage peak periods without needing to (a) ramp up staffing levels, or (b) compromise service levels, but that by doing so also enables restaurants to capture valuable data on their customers that is not usually available through the traditional business model. Thus, the service enables restaurants to build rich databases for purposes such as patron recognition. OpenTable, a US-based company, provides service in Canada, Germany, Japan, Mexico, the UK, and the US, and partners with a host of service providers including Google, TripAdvisor, and Zagat. The service provides online reservations for diners free of charge. Diners may also make reservations via mobile applications available for a variety of smartphone operating systems including Android, BlackBerry, iPhone, and Palm, which is particularly helpful to those who are out and about and looking to try a new restaurant, those who may have just attended a show and are looking for a bite to eat nearby, or those who are traveling. Consumers can use their smartphone's geo-location feature to find restaurants in the vicinity and can check seating availability. One key element of the service is that everything is conducted in real time: search results indicate real-time availability and reservations are immediately reflected in the system. Also, the service allows diners to make reservations at any time of the day or night; they are not restricted to the operating hours of the restaurant.[46]

Zipcar

- Zipcar has shifted the traditional rental car business model to an on-demand model after recognizing that the traditional model did not address unmet needs such as customers only needing cars for a few hours and therefore only making them pay for the hours used, having cars available where customers are (college campuses, downtown areas, and so on) and allowing them to reserve cars via the Web or mobile phone just minutes before they are needed. Zipcar has implemented a myriad of new-generation technologies to serve the customer on the go. In the mobile arena, Zipcar offers a mobile application that may be used on Apple's iPhone and iPod touch to locate and reserve one of the company's vehicles. Furthermore, an iPhone user may unlock and lock the doors and even honk the vehicle's horn via the device. Moreover, other smartphone users in the US may find Zipcars through the GPS mobile application WHERE. If users choose to do so, they may sign up to receive alerts near the end of their reservation. If the car is still available and the user wishes to extend her reservation, she may do so by simply texting back to the alert notification. Members receive a Zipcard which enables them to gain access to the Zipcars (there is a card reader in the windshield). Zipcar leverages RFID technology so that the member is recognized as having a reservation for the car and may gain access to it. The Zipcard gives members 24-hour access to Zipcars, 7 days a week, all over the US, as well as some cities in Canada and the UK. As the company's website highlights, Zipcar continues to work each day to ensure that members may easily access Zipcar's technology.[47]

Takeaways

Despite the numerous constraints faced by airlines (such as their disparate legacy systems and applications), airlines have implemented numerous technology-led achievements such as Web and mobile check-in, virtual assistants on their websites, self-tracking of baggage status, and the beginnings of door-to-door services. While many airlines have already implemented social media in terms of customer service, some airlines have also begun to embrace social media to enhance their brand and sell more tickets. Airlines are pursuing context-aware and location-based mobile applications through the detection of a traveler's location and the delivery of targeted information and services. Airlines can gain significant insights from the experience of other businesses, especially Zipcar, that has implemented a myriad of new-generation technologies to serve the customer on the go.

Notes

1   united.com/traveloptions/
2   "11 innovative airlines to keep an eye on in 2011, www.airlinetrends. com, 25 January 2011.
3   "ATW's 37th Annual Airline Industry Achievement Awards," *Air Transport World*, February 2011, p. 32.
4   Ron Reed, "Get smart," *Airport World*, October–November 2010, pp. 43–4.
5   "11 innovative airlines to keep an eye on in 2011, www.airlinetrends. com, 25 January 2011.
6   "How Mobile Technology Will Enhance Passenger Travel," *New Frontiers Paper*, SITA, 2009.
7   Further information can be found at Next IT's website, www.nextit. com
8   www.siri.com/
9   "New Delta food areas at JFK let passengers order food at the gate via

iPads,"www.airlinetrends.com, 3 December 2010.

10   Gregg Brockway, TripIt, Presentation at the OSU International Airline
     Conference, Seeheim, Germany, May 2010.

11   www.frommers.biz/

12   www.reardencommerce.com/products/

13   American Airlines offers its Five Star Service program at more than
     a dozen airports where passengers receive, for a fee, a VIP airport-
     assistance experience. For more information, see American's In-Flight
     magazine, *Americanway*, June 01, 2011, p. 70.

14   http://www.virgin-atlantic.com/en/us/manageyourflights/
     diycheckin/altcheckin.jsp/

15   united.com/traveloptions/

16   "11 innovative airlines to keep an eye on in 2011, www.airlinetrends.
     com, 25 January 2011.

17   http://www.etihadairways.com/sites/etihad/global/en/
     experienceetihad/attheairport/pages/GuestTransfer.aspx/

18   http://www.virgin-atlantic.com/en/us/whatsonboard/upperclass/
     servicesandextras/index.jsp/

19   "11 innovative airlines to keep an eye on in 2011, www.airlinetrends.
     com, 25 January 2011.

20   Ron Reed, "Get smart," *Airport World*, October–November 2010,
     pp. 43–4.

21   Mary Kirby, "Personal Touch," *Airline Business*, June 2011, pp. 54–6.

22   Lori Ranson, "Airlines Business' The Airline Strategy Awards—Annual
     Marketing Award Presented to JetBlue," *Airline Business*, August 2010,
     p. 36 and Madhu Unnikrishnan and Robert Wall, "All That Twitters:
     JetBlue and Virgin America champion social networking, but legacy
     carriers lag behind," *Aviation Week & Space Technology*, January 11,
     2010, pp. 42–4.

23   Mary Kirby, "Social Status: Airlines from around the globe are
     adopting social media marketing strategies to connect with passengers
     on a deeper level. Their aim: to create brand ambassadors for years to
     come," *Airline Business*, March 2010, pp. 40–2.

24   Lauren Lovelady, "Chatter Box," *Ascend*, 2010 Issue No. 1, p. 29.

25   Michele McDonald, "Social Distribution: Sites like Facebook and
     Twitter have become important communications channels. Can they
     also sell tickets?" *Air Transport World*, November 2010, p. 58.

26   Brian Halligan and Dharmesh Shah, *Inbound Marketing: Get Found
     Using Google, Social Media, and Blogs* (Hoboken, NJ: John Wiley, 2010),
     p. 116.

27   Liana "Li" Evans, *Social Media Marketing : Strategies for Engaging
     in Facebook, Twitter & Other Social Media* (Indianapolis, IN: Que

Publishing, 2010), p. 299.

28   For further information see Kevin H. Nalty, *Beyond Viral: How To Attract Customers, Promote Your Brand, And Make Money With Online Video* (Hoboken, NJ: John Wiley, 2010).

29   Jim Sterne, *Social Media Metrics: How To Measure And Optimize Your Marketing Investment* (Hoboken, NJ: John Wiley, 2010), p.xxviii.

30   Dave Evans, *Social Media Marketing: The Next Generation of Business Engagement* (Indianapolis, IN: Wiley Publishing, 2010), p. 9.

31   Mary Kirby, "Going Viral," *Airline Business*, March 2011, pp. 38–40.

32   Mary Kirby, "Going Viral," *Airline Business*, March 2011, pp. 38–40.

33   Gillian Jenner, "Vital Statistics," *Airline Business*, March 2009, p. 44.

34   "Transforming Airline Operations at Airports Using Handheld Devices," *New Frontiers Paper*, SITA, 2010.

35   "New technologies for reservations & ticketing," *Aircraft Commerce*, December 2009/January 2010, pp. 22–4.

36   Ron Reed, "Get smart," *Airport World*, October–November 2010, pp. 43–4.

37   Andrew Heasley, "Virgin Blue uses BlackBerry app to lure business travellers," www.theage.com.au, February 24, 2010.

38   Rimma Kats, "Expedia takes travel planning to new heights with mobile," www.mobilemarketer.com, March 2, 2010 and "TripAssist Brings the Ease and Convenience of Expedia.com to Millions of Mobile Travelers," *Barrons* (online edition), February 26, 2010.

39   "ATW's 37th Annual Airline Industry Achievement Awards," *Air Transport World*, February 2011, p. 25 and Mary Kirby, "Airlines Business' The Airline Strategy Awards—Annual Technology Award Presented to Emirates," *Airline Business*, August 2010, p. 38.

40   "11 innovative airlines to keep an eye on in 2011, www.airlinetrends. com, 25 January 2011.

41   "Two world firsts for Malaysia Airlines as they harness the iPad and iPhone," *Air Transport News*, 23/06/2010.

42   "11 innovative airlines to keep an eye on in 2011, www.airlinetrends. com, 25 January 2011 and "ATW's 37th Annual Airline Industry Achievement Awards," *Air Transport World*, February 2011, p. 28.

43   "11 innovative airlines to keep an eye on in 2011, www.airlinetrends. com, 25 January 2011.

44   "The 'Digital Aircraft'— Heralding a New Generation of Aircraft Operations," *New Frontiers Paper*, SITA, 2010.

45   "The Future of Service Business Innovation," *Tekes Review* 272/2010, p. 29 and www.ifholdings.com/

46   "The Future of Service Business Innovation," *Tekes Review* 272/2010, pp. 40–1 and www.opentable.com/

47   "The Future of Service Business Innovation," *Tekes Review* 272/2010, pp. 25–6 and www.zipcar.com/

Chapter 4
Potential Information- and Technology-Driven Initiatives

While the previous chapter focused on how some airlines (and some airports) have been deploying evolving technology during the first decade of the twenty-first century, this chapter provides some insights for potential technology-enabled initiatives during the second decade. This theme is centered on engaging with the digital and mobile passenger to introduce improvements in four areas: (1) passenger segmentation, (2) passenger relationship management, (3) passenger loyalty, and (4) a flawless passenger experience, as illustrated in Figure 4.1. These components are, of course, not independent. For instance, better engagements with passengers can lead to better passenger segmentation and more relevant innovations in service (at the individual passenger level), that, in turn, result in greater loyalty. While these components are not new topics, there are new challenges and opportunities within each component, given that passengers have now gone digital and become much more mobile. It is the potential opportunity for some traditional legacy airlines to use their resources (their huge networks, alliances, enormous customer databases) to exploit information and technology to transform from being operations and product centric to customer centric to compete much more effectively with the low cost carriers and the potential entry of high-powered technology businesses in the distribution arena. (While there are potential data privacy issues associated with the collection and use of information, the discussion is outside the scope of this chapter in light of the complexities involved.) The chapter ends with some examples of the best practices from other businesses, such as the hotel citizenM and the low cost bus service provider megabus.com.

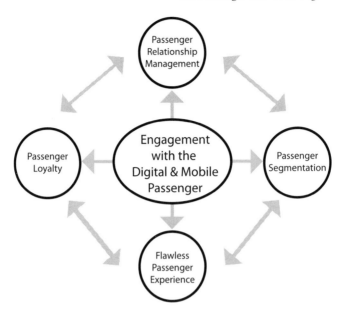

Figure 4.1   Passenger-Centric Initiatives through Passenger Engagement

Passenger Segmentation

Until about the 1970s, relatively speaking, passenger segmentation had evolved at a very slow pace. Airlines segmented passengers typically by trip purpose (business, leisure, VFR), by length of haul (short-, medium-, and long-haul), by region (domestic vs. international), by service (first class vs. economy class and charter vs. scheduled), by itinerary (inter-line vs. intra-line and direct vs. connect), by distribution channel (airline vs. travel agent), by point of sale within an airline (city center ticket office, airport check-in, airline call center), and so forth. Then came the frequent flyer program that enabled airlines to segment passengers by their tier status. The evolution of technology enabled airlines to establish new forms of segments. For example, the introduction of GDSs, followed by the entry of online travel agents, and the technology-enabled proliferation of low cost carriers led legacy airlines to broaden their segmentation process even further. Next, the development of revenue management systems took the segmentation process into a deeper dive when airlines began to segment passengers into much finer sub-segments based on passengers' willingness and ability to pay different fares. Finally, the availability of the Internet (enabling sales through an airline's own website), discovery of ancillary revenue sources (enabling segmentation by core vs. non-core revenue), and mobile and self-service technology (allowing segmentation by the source of check-in) facilitated airlines to further refine the segmentation process.

However, despite these improvements in the segmentation process, and despite the availability of new-generation technology, airlines are still not at the point of targeting *individual* passengers with respect to individual customization. One reason could be that airline reservation systems have been structured around a silo approach in terms of booking engines, inventory, customer relationship management, and pricing. Since these sub-systems do not interface with one another, airline sales channels do not possess the ability to identify individual passengers and offer them fares as well as ancillary products based upon their previous buying behavior and current needs. The challenge could be related to cultural issues in that distribution managers did not push hard the perception that IT issues were cumbersome. Retailers, such as Amazon.com, capitalized upon technology much earlier enabling the company to identify individual customers and target them through offerings based upon their previous purchases. Airline reservation systems suppliers are now beginning to develop the means for airlines to offer products that are utilizing smart and predictive technologies to identify and target individual consumers. For example, one feature is for an airline to be able to offer different fares at the same time to different individuals based upon their individual buying behavior and future expectations. Another possible use of this new technology is the ability to offer different ancillary products and services based upon previous buying behavior (such as some pay extra for aisle seats, while others pay for airport lounge access, and yet others want travel insurance).[1] The next step could be a functionality that allows passengers to make trade-offs in real time. A passenger who is not a member of an airport lounge may choose to give up the mileage credit for a particular flight that has been delayed for a long time in return for the use of a lounge. Then there are passengers who would be willing to provide various levels of ancillary revenue for various levels of guarantees for on-time performance.

The first critical success factor in the segmentation process is to exploit the data that airlines already possess on individual customers by profitability. Up to this point, for the most part, this data has usually been used for frequent flyer programs, and not pursued for other means. As for collecting new data, many airlines do not track and keep basic information about passengers such as records of passengers whose baggage has been lost (by cause), or passengers who missed their connections (this is only done for current day, not historically to determine trends, and, again, not by cause). Such information can be used to communicate with passengers on mutually acceptable solutions, ranging from simple apologies to different types and forms of compensation based on the situation and on the relative importance of the passenger, and based further upon the passenger's previous buying history. However, these passenger-centric strategies cannot be implemented even when data is available due to the different silos of information existing in the airline. The silos do not interface; data is not segmented at the passenger-centric and actionable levels, and there are rules-driven systems and processes. Furthermore, as many airlines consider the information on their passengers as their important assets, they do not share the *relevant details* with their alliance partners. Consequently, alliance partners do not have pertinent and detailed information on the passenger profile or preferences of a partner's passenger that they host on interline or code-share flights.

If the foundation for effective segmentation is the availability of raw data at the personalized level, then what is the source of such data? The mobile channel can now allow airlines to create rich data banks where technology can be used to segment the data into extremely personalized cells, further segmented by situation. However, such levels of individualized marketing require first that airlines disaggregate passengers differently. For example, instead of just business or leisure, passengers could be segmented within those two sub-segments. For business, such classifications could include those who travel overseas, those who travel just within the US, those who travel to one/the same destination every week in the US, etcetera. The business traveler could even be further disaggregated by type of job-related travel. This would be helpful as senior executives traveling for business may have one set of expectations and needs, whereas those in sales traveling to various locations may have different needs. Segmentation can then proceed to the direction of travel. A delay has a different value to a business passenger flying out to attend a meeting versus one returning home having attended a meeting. Similarly, while airlines have segmented passengers by length of haul for decades, what has not been done is the assignment of value for different sub-segments. A two-hour delay to a passenger planning to travel on a one hour flight first thing in the morning and returning in the afternoon is not the same as the same delay to a passenger traveling half way across the world. Similarly, while airlines have been segmenting leisure travelers in standard categories such as students, families, and empty nesters, there needs to be a classification by need. A delay to a student going home for a vacation is not the same as a delay for a passenger who is going to miss a 7-day cruise ship. For personal travel, one passenger could be visiting friends and relatives for a week. A second one could be flying to participate in a wedding or attend a funeral. A lengthy delay or a missed connection has a different meaning and different implications for different passengers at different times.

The key is to understand what motivates each passenger's behavior. In the case of the business traveler's selection of an airline, is it truly his personal preference, or the corporate travel policy mandate? It is understanding the what, why, and when of passenger behavior that the segmentation process needs to address to enable airlines, as well as travel service providers, to implement appropriate customized strategies, based on time, location, and situation. It is also necessary to take into account the fact that a given traveler may have multiple profiles, such as when she travels for business versus when she travels with her family. These different profiles could allow the airline to capitalize on the different needs specific to each scenario. The challenge that airlines face is that many tend to segment on attributes that do not allow them to recognize easily that a traveler in different situations might actually be the same person: in one situation she is the high-value business passenger, yet in another situation she is traveling with the family on low-fare tickets. Therefore, it is not even a segmentation of one, but rather multiple profile segmentation within a segmentation of one that is needed to provide truly individualized service that is also specific for a given situation. The passenger traveling with her family on the weekend described above cannot simply be labeled as an "economy-class passenger" when in fact she is the same person that is paying full fare during the week for business travel.[2] The solution might be to come up with a smart segmentation addressing various segment dimensions. Again, many airlines struggle with making investments in the collection and synthesis of data, let alone in analytics technology for analysis and predictive behavior (in which behavior refers to not just the "what" of travelers' purchases, but also the "how" of the purchasing process, a concept discussed in Chapter 6). Next, even if the data and the analytic techniques were available, airline staff needs to have access to and act upon this multi-profile data to provide the appropriate service accordingly. Within the same cabin class, the same passenger could be interested in a quiet zone one day and an entertainment zone the next day. The multi-profile feature of the segmentation process will not only enhance the on-board experience for a passenger, but also provide an opportunity to increase ancillary revenue, thus resulting in the possibility of increasing the overall revenue yield per passenger.[3] Segmenting the customer as finely as described clearly has a number of challenges. For instance, a traveler may be limited in his capacity to behave as expected, or as he might prefer, if he is limited by his corporate travel policies, as mentioned above. Such limitations are likely to impact the amount of true segmentation that may be feasible.

While the mobile channel is one rich source of data, the traditional website is a second source. The real value may be in tracking not what people on the website searched or even purchased but what they wanted to buy. One example of an airline that has started to focus its strategy on addressing the individual needs of passengers is Lufthansa. The airline introduced, for example, the "trip finder" tool, an Amadeus-developed tool that does not require passengers to indicate a destination city when on the airline's site, but rather gives passengers the flexibility to search based on (1) their budget, and (2) their interests in terms of the types of activities that they wish to engage in during their trip. Such an initiative helps the airline engage customers at the inspiration phase of the travel process, and transform the inspiration into a booking.[4] However, while a good start, such a system only looks at the cost of the air ticket that in many cases can be only a fraction of the total cost of a trip. The additional cost, driven by personal preferences relating to hotel accommodations and restaurants as well as the typical cost of vacation-related activities (skiing, golfing, and so forth) could be added if there were a better profile of a customer. The focus of most airlines continues to be on the limited "airport-to-airport" component of the trip.

Improving the segmentation process within an airline is a first step toward becoming customer centric. However, since many airlines work within an alliance framework, coordination of segmentation profiles with other alliance partners is a major challenge and an opportunity. For example, if a passenger has had his baggage lost twice on Airline X, then an alliance partner, call it Airline Y, needs to know that on this passenger's next flight. It becomes imperative that he is given special treatment in light of his recent experiences. While airlines may not want to share some specific data, general non-competitive information could be shared among alliance partners such as that described above to help ensure more of a consistent, as well as customer-centric, experience among alliance partners. What is the point of being a part of an alliance if each member is going to act as if it is just out for itself? Technology could be leveraged to share this type of data among alliance partners. According to many passengers, alliance partners should do just that, act as a real alliance, for example, offer passengers the best routing for the best fare. If it is going to cost a passenger US$900 to go through a US gateway on a US carrier, but the passenger could travel on an alliance partner through a Canadian gateway for a several hundred dollar savings, then the US partner should inform the passenger of such information. Many passengers perceive that only when an airline is pushed against a wall in times of crisis (delayed or cancelled flights) will it consider putting a passenger on a different carrier, even if the carrier is part of the same alliance. Some alliances do share data on a transactional basis (individual information relating to a specific instance of a passenger need) but not on a broader basis. If more alliance partners were to adopt such practices, it would be an improvement for the whole alliance.

Again, as perceived by many passengers, airlines need to move away from the "keep it on my airline" mentality. Technology and willingness to act as part of a real team, that is, the alliance, can be leveraged to evaluate alternatives that are in the best interest of an *individual* passenger. A truly customer centric alliance could "own" the customer as a group. However, the benefits of technology and data sharing will not be effective unless alliances dramatically change their commercial approaches. This step may be too great for some carriers to take, especially for the smaller carriers who may feel marginalized in the alliance. According to many seasoned travelers, alliances appear to be competition centric as they want to minimize capacity and fare competition within the group with products like code-share, interlining, or new ideas of an integrated schedule pricing system on the North Atlantic. Think about the value of "owning" even the small segment whose highest priority is on-time performance, as discussed below. The challenge is related to the need for common IT platforms within alliances. One way is to have a common core. Another way might be to have a two-class system within an alliance, such as full and basic benefits.

Passenger Relationship Management

Airlines do have valuable data on customers but many do not know where it exists. It could be in call centers, loyalty programs, customer service, CRM, with strategic partners, and so forth. To maximize the value to be derived from the use of social media, there is a need to first integrate the fragmented IT systems and data marts. One airline, considered to be a leader in this area, progressed from having numerous separate customer databases to just one data warehouse. Before, the airline struggled with knowing who its top customers were and their associated value. The new system, rather, involves one data warehouse that contains all customers and their associated data (frequent flier data, travel history, Web history, and so on) not only all in one place but with the ability to access the majority of the data in real time. This allows the carrier to provide frontline employees with relevant data to provide customers with the appropriate service in real time. In the longer term, it allows the airline to understand customers' value and what they value and target them accordingly.[5] It is the combination of integrated information and technology systems that can lead to driving maximum value from the use of social media, not only in terms of additional revenue streams, but also in terms of building the brand.

While some airlines may have taken the initiative to develop a single data warehouse, they may not be using the data to develop a holistic view of a customer, let alone force all departments to use the same data. The number of airlines that are developing sophisticated models that analyze and predict passenger behavior (to identify and create business opportunities that deliver competitive advantage) is even more limited.

Customer Relationship Management (CRM) has been used successfully by some businesses to become more intimate with their customers. However, when considering CRM, it is first important to distinguish between who wants to have a customer relationship with you, and who does not. Consider a bus ride. Most likely, the person is trying to get from A to B, and that is it. It is purely a transaction, and nothing more. Likewise, some passengers may also feel this way about their airline travel. Therefore, it is (a) imperative to segment passengers by those who want to have a relationship with an airline and those who do not, and then (b) to further segment those who do wish to have a relationship by what type of relationship they wish to have. Does a given passenger desire recognition, or purely upgrades, or both? Airlines are conducting CRM, but not to the degree that is needed in this emerging world of personalization. One explanation for why the airline industry has not progressed much relative to other industries may be that the leading users of CRM are probably those selling a broad range of products and services directly to well-identified and well-profiled customers (such as supermarkets and online retailers). Airlines, on the other hand, tend to sell a limited number of products and services, often indirectly through channels that are complex and sometimes opaque to the customer. Consequently, the scope and use of CRM to derive a clear and significant ROI tends to be limited.

The fundamental point is to recognize what is important to a passenger, and this can change in different situations, even though it is the same passenger, as mentioned in the previous section. This level of information can only be achieved by systematically collecting data at all touch points and on an ongoing basis. Digital and mobile technologies as well as social networks can help airlines achieve information with this level of detail. Take the case of social media. It has been growing at a phenomenal rate based on the statistics of the increasing number of Twitter and Facebook users. Marketers are now beginning to explore the use of combining social media and CRM (referred to as "social CRM"), to get closer to passengers (by recognizing and engaging with passengers who are or could be social influencers) and to react on a timelier basis to passenger feedback and emerging trends. However, while the deployment of social CRM initiatives would appear to be beneficial, according to some skeptical management, it would be difficult to build a business case using traditional methods, such as ROI. On the other hand, if an airline is able to resolve an issue online before it gets to a call center or help desk, then there are measurable savings as airlines have some idea of the cost of dealing with passenger complaints through call centers and help desks. Consequently, it would not be an impossible task for building a business case for the deployment of social media.

The starting point, of course, is to drill down in the integrated data warehouse to identify not only passengers who participate in social networks and online communities, but also the subset who are considered to be influencers or high lifetime value passengers (or both). There are many more options to profile passengers but there is another source not to be overlooked, namely, passenger experience. Airlines do capture on-time arrival and baggage issues, but not the "perceived" experience. This cannot be undervalued. They may be able to capture this through social media. However, this will be a small percentage for quite a long time. An airline can then develop a profile on each influencer, the specific social network used, level of activity (downloads, site visits, and so forth), and the degree and type of influence (for example, referrals). Next, the airline would need to identify not only the tools needed to analyze the information (for example, read emails and perform analysis of texts), but also the optimal channel of communication (email, specific fan page, etcetera) with each passenger who is considered to be a social media influencer. Think again back to the small segment that just wanted on-time performance and the airline consistently delivered this service. Think about the influence of passengers in this segment.

The basic problem at most airlines is not the lack of data, but the lack of coordination within different sub-functions of marketing. For example, different groups are responsible for "listening" to conversations in social media, such as those responsible for CRM initiatives, and then there are those responsible for developing marketing and sales initiatives, such as customer service or a refund policy. Even the group that is responsible for "listening" could be different from the group that is "analyzing," for example, identifying a segment that has passengers who could be getting ready to file a complaint. Each one, in turn, is different from the group that actually develops and takes some action. So, even within enlightened airlines, there is no single group with the responsibility for the deployment of holistic social CRM to respond to emerging trends relating to passenger loyalty, product development, and brand management.

It is eye opening to observe the differences between the quality control that an airline uses in some traditional airline-related areas (such as operations and purchasing) compared to the acquisition and use of information on passengers, through traditional sources or through social media. The group responsible for social CRM needs to facilitate coordination between monitoring and measuring "conversation" across the social networks, data collected through traditional sources and social media interactions, the data analyzed in a context-related framework (separate hype from reality to develop insights), business rules, as well as the analytics to improve service consistency and improve the customer experience from end-to-end. This holistic nature of the function also means that the group should also focus on "collaborative engagement" with passengers in that not only must passenger issues be listened to and resolved, but the group gain insights from passengers about new ways to develop and promote an airline's products and services, not to mention bring about premier transparency.[6]

While traditional analytics processes tend to be more historically focused, social media analytics involves processes that are more focused on real time, as well as other attributes including collaboration. Furthermore, since social media tends to involve conversations (resulting in mainly unstructured data as discussed in Chapter 2), it does not translate well into traditional customer analytics tools. Due to the nature of social media, customer analytics has fundamentally changed: it is necessary for companies to obtain new methods of pulling the data, gaining insights, and acting upon them. In addition to new technologies, different types of analytical skills will also be needed. For instance, agility is a key characteristic in dealing with social media as speed is such a big part of social media: data changes quickly, thus the focus of analysis can change quickly, and in turn, this can affect decision-making processes. Collaboration is another key characteristic in dealing with social media, as learning about customer insights and taking action in this medium involves interacting with customers rather than just managing them, which is a dramatic shift for some firms. Due to scarce resources and high costs, it is most beneficial to create a centralized social CRM and customer analytics function.[7]

It is ironic in that it was the airline industry that was a CRM leader in the 1980s when airlines introduced frequent flyer programs as a means of increasing loyalty. However, while many other industries currently use CRM to identify, grow, and retain valuable customers, many airlines have not moved much beyond frequent flyer programs to capture the true potential of CRM. They simply have not invested in building and leveraging customer knowledge to obtain deeper insights about their passengers. Consider the following statistic: only approximately 40 percent of customers participate in frequent flyer programs, and if many airlines have done little, as it appears, to move beyond these programs, then the remaining 60 percent of customers are therefore treated as unknowns, as little or nothing is captured on these passengers.[8]

Looking at a more general view of analytics, consider the findings of a study reported in the *MIT Sloan Management Review*. First, there are many practical uses of analytics outside of the traditional finance and operations functions. For example, some retailers have leveraged analytics to strengthen their competitive advantage in areas such as customer experience management. Second, for the organization to embrace the new insights derived from the deployment of new analytics they must be (1) aligned with the business strategy, (2) relatively straightforward for all employees to comprehend, and (3) actionable. Third, the study showed that there tend to be three levels at which analytics can be adopted: aspirational, experienced, and transformed. Businesses that have a clear understanding of where they fall within this spectrum may be in a better position to transform challenges into opportunities. Finally, in the near future, the authors conclude that one trend will be to enhance traditional analytics activities with new emerging techniques and tools (such as data visualization) that would yield valuable results in terms of understandable and actionable insights. For an in-depth description of this study, see the report by Steve Lavalle *et al.* in the Winter 2011 edition of the *MIT Sloan Management Review*.[9]

Passenger Loyalty

The first essential element in loyalty is the need for focus on getting the basics right first, which has been the key ingredient in the success of Southwest Airlines. See Figure 4.2. The fundamentals of the airline business continue to include getting passengers from A to B safely, on time, with their luggage, with reasonable service and at a reasonable price. However, often times these operational issues get downplayed as airlines get caught up in focusing on the "extras" or "nice to haves," rather than the basics. According to some passengers, how important is it to dwell on a huge or rare selection of wines available on-board or the hundreds of movie options available to a passenger instead of concentrating on the core elements of an airline's offering, such as on-time performance, particularly for connecting passengers when there is no next flight that day or there are no seats available on the next flight? Furthermore, getting back to the basics also includes focusing on the actual product itself for the largest segment of travelers (passengers in the economy cabin), the airline seat. Examples of airlines that have excelled in this area include Cathay Pacific and Air New Zealand. A few airlines are actually recognizing that operational performance can be a competitive differentiator.[10] Air New Zealand has been dedicated for years on offering excellent coach seats on its long-haul flights. Currently, the carrier is featuring several unique offerings in an attempt to better meet the needs of its economy passengers. These include the ability to purchase three seats and convert them into a flat surface (see photos in Chapter 6), as well as hard shell seats that can be configured in different ways to cater to passengers' needs such as privacy for those traveling alone or dining for those who wish to eat together.[11] In the hotel industry, Westin has focused on its core offering: the bed. The hotel is famous for its "Heavenly" bed, which is even available now in stores, so that guests may replicate the experience in their own homes.

The second element in developing and maintaining loyalty is to keep passengers informed of potential problems and proposing meaningful solutions, as illustrated in Figure 4.2. If a flight is delayed, let all passengers know (not just those with elite status in the frequent flyer program) of the nature and extent of the delay and the potential problems for each and every passenger as well as the proposed solutions with options. If a connecting passenger is likely to miss his connection, what are the options proposed by the airline? The airline could offer a seat on its own next flight (providing a time of departure), offer a seat on an alliance partner (including the routing), a seat on a competitor (including additional fare, if any), and, if appropriate, transportation by an alternative mode. The passenger wants to be kept informed of the status of the problem and allowed to select an option provided by the airline. Moreover, passengers now want to have the ability to serve themselves, beyond the standard self-check-in (including baggage) capability. This passenger requirement presents a different set of challenges in terms of leveraging enabling technology. Again, respond to every passenger. If the airline cannot respond to the passengers on deeply-discounted fares, then those passengers must be informed at the time of their purchase that they will be on their own to solve their problems should disruptions arise, or the airline should not sell to that segment of the marketplace.

The third element in the loyalty triangle illustrated in Figure 4.2 is the development and maintenance of credibility. With the proliferation of information and channels through which it is available, airline marketers are facing a significant challenge relating to the credibility of the information they provide. The full spectrum of checks on credibility ranges from the source of information being relatives and trusted friends (especially if the source is word-of-mouth), to travel agents (brick-and-mortar and online), to random bloggers posting their opinions within the social network and online communities. Passengers now have different sources to compare the claims made by airlines in their advertisements or their websites.

Figure 4.2   Fundamental Elements of Passenger Loyalty

What is an effective tool that airlines can leverage to measure loyalty? Voice of the customer would be one. If passengers feel that they are not being listened to and being responded to, they will go elsewhere. Truly listening to passengers, not just paying lip service, can be important not only in increasing passenger loyalty, but also beneficial in terms of gaining valuable insights relating to product development and service enhancement. While some airlines have deployed the voice of the passenger tool to evaluate their products and the loyalty of their passengers, they are not using it comprehensively to capture its full potential. For example, even the airlines that have implemented voice of the passenger programs, are they providing the resources to closely monitor the quality of their communications with target passengers and to determine if their passengers feel that the airline is responding to their concerns? Problems relate to insufficient information and/or the existence of silos in the airline organizations. The solution, of course, needs to be the dispersal of this tool throughout the airline. In today's environment, the voice of the passenger can be a strategic objective of an airline and every department should be held accountable, not just customer service or even marketing as a whole.

While listening to the passenger is not a new concept for some airlines, there is a need even for them to introduce some improvements. This was not a priority for many airlines as they were focused on cost reduction and survival. As such, there was significant oscillation on passenger satisfaction. Assuming that the right questions are asked, the data tends to be displayed in graphs and tables and rarely made accessible to relevant frontline staff. Even in those cases when it is provided to the relevant staff, it is provided as raw data, with little or no help in relation to the interpretation, let alone the necessary empowerment to respond, based on the location of the touch point. First, it needs to be an ongoing process as passengers' preferences are changing. Second, with the proliferation of the channels of communication, airlines need to interact with passengers through channels of their choice, for example, call centers, emails, IVR (Interactive Voice Response), social network conversations, and so forth. Third, there is a need for the integration of passenger feedback received from multiple channels (and various touch points) to obtain a comprehensive understanding of a passenger's perspective of the service purchased and the service delivered. Fourth, there is a need for timely response. Fifth, there is a need for a follow-up. Finally, there is a need for senior management to connect the voice of passenger initiatives with corporate strategies, particularly for support initiatives.[12]

Airlines can leverage emerging technologies to not only address the three elements of loyalty shown in Figure 4.2, but also to use the voice of the customer tool to measure loyalty. Text analytics and IVR are just two examples of emerging technology that may be implemented to capitalize on the voice of the passenger. Interactive surveys can allow airlines to capture passenger feedback right at the relevant touch point, thus providing truly meaningful and actionable data in real time. Moreover, technology is also available not only to listen and analyze the voice of the passenger, but also to "predict" outcomes of various actions. In the final analysis, it is listening to and acting upon the voice of the passenger that will lead to an increase in passenger loyalty from an improvement in passenger experience.

Flawless Passenger Experience

Customer service is a very broad function. The thread that weaves through various elements of customer service is passenger experience, that may relate to a myriad of areas including shopping and buying air travel, checking-in and flying on an airplane, or dealing with problems that may incur en route (such as delays and cancellations). Although it is this passenger experience and its flawless execution that can be facilitated by new-generation technology, the critical success factor in the implementation of the relevant technology is a change in the corporate organizational structure and culture. First, airlines need to implement strategies that focus on information-based personalization. Personalization can help regain control of passengers that some airlines have slowly lost over the years. Passengers want to be treated on an individual basis, not in broad segments such as business, leisure, and visiting friends and relatives. They do not even want to be treated as the same within sub-segments such as those belonging to a particular tier status. As stated numerous times throughout the book, not only is one passenger different from another, but the same passenger's needs can and will vary by situation. Second, the strategy should be developed within a holistic framework. An airline might be able to acquire the technology, but the staff to make it happen is just as important. In many cases cabin crew is the interface. Third, the strategy should be developed from a contemporary and best practices view point, not a historic or an airline-industry-specific view point. These three points are illustrated in the following discussion with respect to mobile passenger-enabling initiatives.

The expansion of the Internet makes relevance an important differentiator as passengers are bombarded with information. Airlines that can leverage information and technology to deliver the value-added aspect of personalization are likely to be successful, as passengers continue to demand individualized services to improve their experience. Mobile devices are a means of delivering personalization with information that is relevant and timely, providing a passenger with "solution" options. Passengers are now becoming more and more digital, mobile, and connected. Connectivity can be leveraged to provide value to the passenger during times of both regular as well as irregular operations, but also to the airline in the form of profits generated by value-added services enabled by this technology. Consequently, airlines should not only start to implement a mobile strategy, but also they should do so in a holistic and contemporary framework, serving as one part of an integrated set of channels (mobile, traditional PC, tablet PC, kiosk, and so forth). The services and products offered can be differentiated in these channels but should still follow a consistent and holistic strategy.

One potential application of the mobile device in air travel would be the ability to sell empty seats to passengers at the last minute so that they may ensure that they will not have to sit next to another passenger. If the average load factor is approximately 80 percent, the airline can capitalize on the empty 20 percent, not in terms of selling those seats to more passengers, but rather in selling them to existing customers who wish to have more personal space. Moreover, passengers could even be asked to actually bid for these extra seats, a process that would unveil the true value of the offering. Technology can be leveraged to enable such an offering to passengers. The potential revenue that such an offering could generate would most likely be proportionate to the length of the flight. For example, a passenger might be willing to pay US$20 extra dollars to have an empty seat next to him on a 2-hour flight, but perhaps US$40 for a 4-hour flight, and possibly even US$100 for a 10-hour flight. The notion of selling spare seats does raise a question. This question relates to the complexity and variability of the air travel process and the restricted visibility of information in its underlying systems (such as during the late handover of information/control from reservations to departure control systems in many airlines). These issues can be real challenges in attempts to improve customer service and profitability. It is also necessary to rethink many complex processes to enable a simple technology-driven change to take place. In addition, could this initiative limit the flexibility of cabin crew to re-accommodate passengers before the flight to address last minute personal requests? Technology companies are beginning to address these challenges in the development of their new generation passenger service systems.

This potential offering is also a good example of the airline focusing on the basics: the seat and comfort. Air New Zealand is already pursuing the concept of selling an extra seat on select flights, but currently this is done at the time of check-in and for a set price, as shown on the carrier's website, rather than the potential bidding concept via mobile as described above.[13] Moreover, in March 2011, Air New Zealand started its "Grabaseat" marketing initiative that involves a reverse auction. In this concept, the airline begins to offer a seat at the regular price, but then reduces the price gradually until the first bidder purchases it.

A survey by SITA reveals that 66 percent of US business passengers carry a smartphone.[14] As with companies outside the airline industry, some airlines have launched mobile strategies[15] to capture the opportunities of this unique channel. As mentioned above, mobile devices can not only provide opportunities relating to customer recovery such as the opportunity of knowing if a passenger has missed his connection and therefore inviting him to the airline's club, but also an opportunity to enhance ancillary revenue. Go back, for a moment to the concept of selling the empty seats to existing passengers rather than additional passengers. Let us assume that an airline operating with an average load factor of 80 percent can only sell the empty seats at a rate of 10 percent of an average ticket price. Even at this rate the airline could add two percentage points to its profitability.

Most airlines have been implementing piecemeal mobile strategies. For example, while enabling passengers to check-in via mobile devices or even receive a 2-D boarding pass via a mobile device is progress, these are still pieces of a mobile strategy. Mobile strategies need to reach passengers at each of their touch points throughout the travel cycle, touch points that are depicted in Figure 3.1 in Chapter 3. Mobile is also much more focused; the goal should be to target the segment of an airline's customers that would use this application rather than targeting all of an airline's passengers. To be successful in the mobile distribution channel, an airline needs to take a systematic and a holistic approach to implementing its mobile strategy. It does not make sense to initiate an iPhone application or texting to augment sales through, for example, mobile coupon offerings, without a detailed analysis. It makes more sense to think about the sub-segment of passengers to access through the mobile channel and how to connect with them. Only after considering these mobile challenges and opportunities is it then appropriate to consider which technologies to utilize (such as iPhone applications and text messaging). Finally, personalized marketing should not just occur at the point-of-sale, but should also carry through and occur at the point of delivery.

Let us consider the customer factor. Mobile strategies are less successful if airlines do not take the time and effort to understand how their passengers utilize their mobile devices. Consider, for a moment, the spectrum of mobile phone users, ranging from a small segment that uses the device to make calls for emergency purposes to a segment that uses it virtually 24/7 for calls, email, Web surfing, social networking, and entertainment. Next, an airline needs to decide what it is that its mobile initiative should accomplish. Within this objective setting step, there are a few points to consider. First, what is the overall business strategy of the airline? Is the emphasis on reducing costs, or increasing revenue, or improving customer experience? Second, the airline needs to identify the overall business objectives that support the business strategy. For example, increasing revenue may be accomplished by either having existing passengers travel more, or by acquiring new passengers. It could be trying to sell the empty seats to new passengers or sell them to passengers already booked on the flight. Third, an airline must choose relevant mobile objectives that support the business goals. For example, a number of airlines are striving to improve customer experience by offering mobile check-in and boarding-pass services.

Next, the airline can design a plan of action to meet the above objectives. Here are some questions to consider in the formulation of mobile strategy:[16]

- How many of its customers does the airline want to reach and the resources required reaching this group?
- What will the airline's mobile offering be within this channel (information, services, or product features)? Mobile experiences are unique and need to capture the opportunities that location and timely delivery offer.
- How will the airline deliver its mobile offering? Should an airline distribute an application directly, via its alliance, or should it be through a third party?
- What is the airline's level of commitment to mobile? Again, consider the resources required. For example, will an additional service center need to be staffed to support the offering?

Once an airline has gained an understanding of its mobile passengers and its own objectives and strategy, it is then appropriate to consider what mobile technologies to implement. Again, it is necessary to conduct research such as the percentage of mobile users that use the device for simply messaging, the percentage that use the device for accessing the Web, those who do both, and so forth. In addition to the percentage breakdown, it is also important to have information on the usage rate. Mobile technologies must be adopted carefully, as the information researched must relate to travel. Finding out that half of mobile users are using their devices to send and receive text messages does not necessarily mean that the airline's customers fall into this group. Similarly, implementing an iPhone application when the majority of an airline's customers use a different device will achieve little success.

These are common mistakes which can be avoided by taking the time to conduct a segmented profile of an airline's passengers. It is quite possible, for example, that 50 percent of an airline's passengers may use the texting and email features on their mobile devices and 25 percent may use their mobile devices extensively for everything (from mobile Internet, to mobile social networks, to entertainment), both numbers being much higher than the national average.[17] Moreover, while only a tiny percentage of the national population may use the mobile devices to check flight status, the percentage may be much higher for a particular airline flying business markets. Furthermore, there must be consideration of the rising tablet PC market as well. In addition to Apple's iPad, other competitors are rushing to release their version of tablet PC (at various levels in terms of price and functionality). The takeaway for airlines is that at one end, consumers are obtaining and utilizing much more expensive and complex devices, while at the other end people are choosing less expensive options with less features. It is necessary to meet the needs of passengers at both ends of the spectrum.

For different segments, the objective can be quite different in implementing various combinations of voice, text, and mobile Internet to reach all of an airline's passengers. Airlines have already been utilizing text alerts for flight updates and status changes. More recently, they have begun to provide boarding passes through mobile devices (subject to the approval of security staff) and to offer a better customer experience relating to airport processes and procedures. While airlines do possess the ability to reach almost all of its passenger base through voice, texting, and/ or mobile Internet, passengers can be given the choice of signing up for text or email alerts when they book their flights. The key goal is still to not only provide information that is relevant and in real time, but that is accurate. When a flight is delayed, an airline should use all the information and technology that it has to estimate how late the flight is likely to be based on the status of the aircraft and its crew relating to all the prior segments. It is ironic that some external organizations claim that they are better equipped to estimate the delay more accurately than the airline's own staff at an airport (See, for example, the discussion on FlightCaster.com in Chapter 2).

Coming back to the holistic framework aspect in formulating a mobile strategy, airlines cannot lose sight of the branding factor. As consumers become increasingly connected (changing their lifestyles to adapt to the digital and mobile world), the first wave of "new" distribution channels (websites) and "new" promotional initiatives (emails and standard banner ads) are becoming obsolete, making the brands, in turn, less effective. To build effective and relevant brands airlines can use step-changing consumer technologies to stay abreast with not only the digital and mobile lifestyles of their passengers, but also those segments who are joining communities of interest and seeking experience that is personalized, compelling, and in real time. Consequently, advertising campaigns that are just location-based (available on mobile devices, for example) are not sufficient. They must also be interactive and engaging. Consider, for example, an ad banner that appears next to content that a passenger has accessed through a smartphone. The ad banner must not only adapt to the behavior of the user, but also must enable the user to link to other relevant micro-sites that go beyond the standard text-based and menu-driven websites.

One example of a global airline that has had initial success in launching a mobile strategy aimed at providing seamless service to passengers is Malaysia Airlines. The carrier did not simply replicate its website for mobile devices. Instead, the airline created a mobile offering that is simplistic; the platform involves one Web application accessible by just one click on a single link. Moreover, it does not involve any downloads to the device. The mobile site, flymas.mobi (developed by SITA Lab), enables the airline's passengers to book and hold flights, purchase tickets, check-in, and receive 2-D boarding passes. Passengers can also view the carrier's schedules and check flight status. In addition, if a passenger's bag is lost, he can report it as well as trace it via the carrier's mobile site. Finally, the airline offers travelers special deals through the mobile channel. Malaysia Airlines' flymas. mobi works with all Web-enabled mobile devices including BlackBerry, iPhone, and Android Google devices.[18]

What does the future hold in terms of mobile and travel? The next step that is already in place involves the bar-coded boarding pass (BCBP) becoming available on mobile devices. This practice will augment the passenger experience, enabled by Near Field Communication technology (NFC). Passengers will receive their boarding information on their mobile devices, information that can, in turn, can be relayed to appropriate readers/sensors as the passenger progresses through the airport.[19]

Stepping back and looking at the whole travel cycle, many possibilities have been identified for both airlines and airports to leverage mobile devices, technology and applications to truly personalize service in the future, as highlighted by SITA. One possibility that has been identified is the use of mobile applications that leverage the GPS function of the mobile device to assist travelers in areas such as finding the shuttle bus, finding a parking space, and so on. Similarly, location sensing technology through mobile devices could assist airlines in managing passenger flow in terms of determining the location of late passengers as final boarding approaches. Another opportunity is leveraging mobile devices to shorten processing times when moving through the airport. Smartphones could be used to store information (visa information, frequent flyer card information, baggage information, even credit card information) as well as to gain access (such as to lounges) thus augmenting the travel experience as it would keep travelers from having to carry so many paper documents and plastic cards. In connection with the concept of the smartphone containing a traveler's baggage information (in the form of an electronic receipt) is the possibility of offering reporting and tracking services for those who have suffered lost luggage. The idea would be that the data contained in the electronic receipt could be utilized in the reporting tool, ultimately making the process more efficient. Further advances in terms of baggage could include travelers having an RFID chip in their baggage that would work in conjunction with travelers' mobile devices; an application could let a passenger know that his bag has arrived, thus alleviating the need to constantly search the carousel. In looking toward the future of personal travel concierges, it is interesting to note the prediction of the transition from a place where travelers interact with the applications on their mobile devices, to a place where smartphones could autonomously make recommendations. For instance, while in the traveler's pocket, the smartphone, equipped with advanced sensor interfaces, could possess the capability to collect data (based on factors such as time, position, and travel plans) and provide relevant solutions such as whether it would be more prudent to take a taxi or a train at a given time to the airport.[20]

Furthermore, consider, as illustrated by SITA, how mobile could be leveraged to improve ground operations through integration and efficiency by equipping staff with the correct information, at the correct time, and at the correct place. The shift from voice communications and paper forms to data-centric mobile devices could yield a myriad of benefits, in addition to the those listed above, namely the timely flow of relevant data to all involved, including a collaborative environment, the transition from a "blame" culture to one that fosters continuous process improvement, and the collection of event data to conduct more substantial cause and effect analysis. Furthermore, the transition to data-centric mobile devices equips staff with "intelligent" applications such as context-aware applications that can ascertain where not only employees are, but also equipment, thus ensuring that not only are the correct people equipped with the correct information, but that they also have the correct equipment, all at the correct place and time.[21]

In conclusion, in terms of business strategy facilitated by new-generation technology, an airline should focus on a combination of seeking opportunities to grow revenue, reduce costs, and improve passenger experience, a balance that any business would strive to achieve. The airline industry is highly competitive and is challenged by the fact that (a) its operations are highly dependent on the weather, and (b) while a significant percentage of airline passengers may say that they would pay more for better service, they, in fact, choose an airline based on the lowest price. New-generation technology can help airlines overcome, to some extent, these challenges by not only raising the bar in providing a higher level of service, but segmenting the passenger base with respect to coming pretty close to truly customized price-service options. It is in this area that the well-established and technology-savvy airlines can leverage their resources to compete with many low cost carriers. It is important not to overlook that flawless passenger experience is closely linked to the integration of functions within an airline as well as an alliance to provide an authentic seamless passenger experience.

Examples of Best Global Business Practices

One example of a best global business practice would be BMW as described by Dean Marci (of the mobile marketing firm, the Cielo Group).[22] Suppose a person is looking at a location-based ad (say a billboard in the gate area of an airport). The ad could contain a key word and a short code. A person on a mobile phone could type in the keyword/short code (say the three letter code of the airport, such as bmw.lax), an action that would take the user to a page that would transmit a video or a photo gallery, depending on the functionality of the receiving mobile phone. The user could even enter a zip code and be given the list of dealers in that area, not to mention the ability to communicate with a dealer and book a time for a test drive. The business using such marketing initiatives would not only know the number of people who took a serious interest in the ad, but also provide the contact information of the user.

A second example of a best global business practice is citizenM, a low cost, high tech, trendy hotel which targets the modern traveler. The company has determined the new traveler to have such characteristics as is an explorer, has a respect for different cultures, is independent, is a professional, and is young at heart. In fact, the "M" in its name stands for "mobile" and the company refers to this new traveler as a "Mobile Citizen of the World." The hotel chain believes that this mobile citizen likes great value, entertainment, sociable atmosphere, and stylish design. The value proposition that the hotel offers is a unique, quality experience at an affordable price (rates begin at 79 Euros per night). One feature of the citizenM that stands it apart from other offerings is its use of technology in the check-in process. A guest utilizes a touch-screen kiosk to check-in by typing her name, email address, or reservation number (also, the company only accepts reservations online). An "ambassador" is available on site if the guest needs assistance, but otherwise she does not interact with a typical front desk. The technology does not stop there, however. In the room, a hand-held device controls the lighting, television, temperature, and window blinds. The hotel features complementary Wi-Fi, and also offers Apple iMacs in the lobby for guests. The hotel is minimalist: there is no gym and no room service. However, a guest can pre-order breakfast at check-in, and then pick it up at the lobby bar. An "ambassador" is available to make coffee or tea to complement the paper bag breakfast. In addition, a self-service area, complete with microwaves, is available 24 hours a day for drinks, sandwiches, and pre-packaged meals. The first citizenM was launched in Amsterdam at the airport, followed by another one in the city. There are plans for expansion across Europe and other parts of the world.[23]

A third example of a best global business practice is megabus. com, a low cost bus service provider that is a game changer in its industry. Unlike traditional bus carriers, tickets are sold online and there are no bus depots. Travelers book online, receive a reservation number when booking, and then use that number to board the bus. The fleet consists of single and double-decker buses which feature power outlets and free Wi-Fi service. The company utilizes a pricing structure that is similar to that of discount airlines: a limited number of seats are featured at just US$1, and then the fares increase according to the time between the booking date versus departure date. Currently, the Chicago-based company serves almost 50 cities in the Northeast and Midwest regions of the US and Canada, with hubs in Chicago, New York, Philadelphia, Washington DC, and Toronto.[24] The key point is that megabus.com uses technology not only to interact differently with its customers, but it really changes the traditional business model standardized by Greyhound. With its sidewalk stops it saves the cost of maintaining bus stations and, at the same time, gets closer to its customers.

Even some taxi cabs in Tokyo are providing free Wi-Fi. Airlines and airports should offer Wi-Fi in every terminal, just to ease communications with passengers as well as enabling their navigation/tracking. How often does it happen that a connecting passenger does not make it to the departing flight on time, which requires his baggage to be off-loaded. Then, while the bags are searched in the cargo hold, the passenger arrives (close to having a heart attack from rushing) at the gate learning that the jet way has already been disconnected and that there is some trouble with the baggage before the plane can leave the position. Had the airline known that this poor traveler was just around the corner, the guest would have made it and the delay would have been just a few minutes instead of a much longer period of time.

These examples are provided for some airlines to overcome their inward and historic orientation. The need to take on an outward and best-practice orientation is being demanded by passengers who experience better services from non-airline businesses. As of now, with the exception of few airlines, the typical comment of a passenger would be: "This is an airline. What do you expect?"

Who Are the Next Passengers?

A new generation is emerging that Friedrich, Peterson, and Koster refer to as "Generation C" due to their attributes, namely "connected, communicating, content-centric, computerized, community-oriented, always clicking."[25] In general, this group was born after 1990. Interestingly, these authors observe that this group:

- will comprise 40 percent of the population in the US, Europe, and the BRIC countries, and 10 percent of the rest of the world by 2020
- is the first generation that has been exposed to the Internet, mobile devices, and social networking for their whole life

Furthermore, these authors point out that consumers will continue to increase their consumption of digital information, a part of which is not verified. Much of this information is "unanalyzed and unanalyzable—but it will soon be put to material economic use."[26]

What is Next in Air Travel Distribution?

The distribution of air travel is at cross roads. On the one hand, airlines would like to reduce their distribution costs, enhance their ancillary revenue, and not only maintain but increase the direct access to the customer. On the other hand, there are powerful technology companies with an interest to enter the air travel distribution arena by providing better shopping and travel experience. In the case of the former, for example, some airlines started serious discussions with distribution intermediaries toward the end of 2010 about reduction in distribution costs. In the case of the later, Google agreed to purchase ITA, an Internet-based technology company offering air travel shopping and reservation functionalities. What could such an acquisition by Google mean for the travel industry?

It could lead to the reversal of the current trend of unbundling of the airline fare and rise of ancillary products. Google's ability to index and analyze the entire Web[27] could lead to a dramatic change in fare search and the payment structure, as Google could incorporate all additional fees in its results. From one perspective Google could provide air fares customized to the traveler's actual needs. From another perspective, Google's entrance into the distribution arena could lead to an increase in the airlines' marketing costs to compete for paid screen positions in Google's search results.

Historically, many airlines have not made good use of the data available to them. They may not have relevant and comprehensive data since data tends to reside in different silos. One system is used for reservations, while another system is used for frequent flyer programs, and so on. Moreover, these systems are usually only used for the operations specific to the airline, rather than for the greater purpose of the value of the data that each holds. Google, on the other hand, is at the other end of the spectrum, with its ability to, first, capture and store a wide array of relevant and timely data, second, to turn data into vital information (business intelligence), and, finally, to control information. Could Google combine various aspects of the travel process using its enormous capabilities relating to the strategic deployment of information and its abilities to collaborate with the relevant organizations to develop, offer, and sell customized and personalized travel products and services and to deliver a step-changing travel experience across the travel chain? Think about a potential "trip finder" that could be developed through the modification of search engines that combines profiles and preferences of the user with offers already available on the Web. Scope, scale, and convenience would outsmart any product that an airline could offer if it continues to focus on only the air travel component and, moreover, on the airport-to-airport part of the trip.

Google does not have to sell airline tickets, it could simply focus on building truly innovative travel search engines, compatible infrastructure (for example, technology platforms that enable smartphones developed by a broad spectrum of manufacturers to carry an equally broad spectrum of applications), and radically different payment schemes. In doing so, Google, unlike the airlines, could capture valuable information that would provide it with captivating marketing opportunities. While airlines have enjoyed some success in terms of ancillary revenue, especially from those associated with baggage fees, Google could surpass the airlines in such endeavors from the utilization of systems that are built with data warehousing, mining, analytical, and predictive functionalities as a key priority? Moreover, while social media sites (such as Facebook and Twitter) could also enter the travel business (using their huge databases and followers), imagine the potential opportunities if one or more of these social websites were to collaborate with Google. Stretch the imagination further with the potential entry of companies such as Apple with its iTravel application, enabling the passenger to take full control of the travel process, except for the part of flying in the airplane itself. On the other hand, imagine if some large traditional airlines were to synthesize their powerful databases (individually and collectively through their alliance partners) and deploy the full power of emerging information and technology. They could then become truly efficient and effective competitors. The whole topic of distribution is quite complex. For example, there are questions such as: What about the regulation of the display of airline data? How would the technology company access the raw data in relation to the data that the technology company is currently collecting and indexing? Would the technology company work independently or collaborate with airlines and GDSs?

Takeaways

Information can now be leveraged to develop much finer levels of passenger segmentation, passenger relationship management, passenger loyalty, predicted passenger behavior, and passenger experience. The challenge and opportunity is to ascertain what motivates each passenger's behavior. It is this understanding of the "what, why, and when" of passenger behavior that the segmentation process needs to address to enable airlines to implement *cost-effective* customized strategies with respect to such factors as time, location, and situation. Enabling technology, especially digital and mobile, can be leveraged to simplify, streamline, as well as personalize touch points in the travel process in the future, as highlighted by SITA. Two potential areas relate to security clearance and border control. Furthermore, airlines can gain significant insights from the experience of other businesses, especially megabus.com that has implemented a low cost bus service and has become a game changer in its business sector. Emerging generations, such as "Generation C," as well as new entrants, such as Google (through ITA) could also be game changers in the travel service industry.

Notes

1   "New technologies for reservations & ticketing," *Aircraft Commerce*, December 2009/January 2010, pp. 22–4.
2   Anja Wickert, Accenture, Presentation at the OSU International Airline Conference, Seeheim, Germany, May 2010 and Wayne Miller, director travel at global technology company NCR, as reported in an article in *Airline Business*: Gillian Jenner, "Vital Statistics," *Airline Business*, March 2009, p. 43.
3   "The Travel Gold Rush 2020: Pioneering growth and profitability trends in the travel sector," a report by Oxford Economics and Amadeus, p. 21.
4   Julia Sattel, Amadeus, Presentation at the OSU International Airline Conference, Seeheim, Germany, May 2010 and Marcus Casey, Lufthansa head of e-commerce, as reported in an article in *Airline Business*: Brendan Sobie, "Attention Deficit: Can airlines still turn to the Internet to differentiate and gain a competitive advantage, or are all airline websites starting to look the same?" *Airline Business*, March 2010, p. 37.
5   Mike Gorman, managing director of customer technology solutions at Continental Airlines, as reported in an article in *Airline Business*: Gillian Jenner, "Vital Statistics," *Airline Business*, March 2009, p. 44.

6   Based upon various articles by various companies such as Acxiom, Attensity, and eGain in "Measuring the Returns From Social Media," *1 to 1 media*, a division of Peppers & Rogers Group and for more related information see R "Ray" Wang and Jeremiah Owyang, "Social CRM: The New Rules of Relationship Management," Altimeter Group, March 5, 2010.

7   Rayid Ghani and Sarah Bentley, "Integrating Social CRM Insights into the Customer Analytics Function," in *The Social Media Management Handbook: Everything You Need to Know to Get Social Media Working in Your Business*, edited by Nick Smith and Robert Wollan with Catherine Zhou (Hoboken, NJ: John Wiley, 2011), pp. 91–119.

8   Anja Wickert, Accenture, Presentation at the OSU International Airline Conference, Seeheim, Germany, May 2010.

9   Michael S. Hopkins, "Big Data, Analytics and the Path From Insights to Value," *MIT Sloan Management Review*, Winter 2011, pp. 21–2 and associated article/report by Steve Lavalle, Eric Lesser, Rebecca Shockley, Michael S. Hopkins, and Nina Kruschwitz, "Special Report: Analytics and the New Path to Value," *MIT Sloan Management Review*, Winter 2011, pp. 22–32.

10   Roger Niven, "Operational management: the aim is 'unremarkable service'," *Aviation Strategy*, October 2010, p. 15.

11   "11 innovative airlines to keep an eye on in 2011, www.airlinetrends. com, 25 January 2011.

12   "Capturing the Cross-Channel Customer: Use Dynamic Customer Engagement to drive revenue across voice, Web and mobile channels," Genesys and Peppers & Rogers Group, 2010.

13   "11 innovative airlines to keep an eye on in 2011, www.airlinetrends. com, 25 January 2011 and http://www.airnewzealand.co.nz/buy-a-twin-seat-at-check-in/

14   *Air Transport IT Review*, SITA, Issue 2 2011, p. 6.

15   Based upon the concepts in a report by Forrester Research on mobile strategy: Julie A. Ask and Charles S. Golvin, "The POST Method: A Systematic Approach To Mobile Strategy," Forrester Research, April 9, 2009.

16   Based upon the concepts in a report by Forrester Research on mobile strategy: Julie A. Ask and Charles S. Golvin, "The POST Method: A Systematic Approach To Mobile Strategy," Forrester Research, April 9, 2009.

17   Based upon the concepts in a report by Forrester Research on mobile strategy: Julie A. Ask and Charles S. Golvin, "The POST Method: A Systematic Approach To Mobile Strategy," Forrester Research, April 9, 2009.

18   *Air Transport IT Review*, SITA, Issue 1 2010, p. 15.

19   "Scanning the Horizon," *Airlines International*, Feb–Mar 2011, p. 30.

20   "Five Steps of Air Travel That Smartphones Will Change by 2020," *New Frontiers Paper*, SITA, 2010.

21   "Transforming Airline Operations at Airports Using Handheld Devices," *New Frontiers Paper*, SITA, 2010.

22   Rick Mathieson, *The On-Demand Brand: 10 Rules for Digital Marketing Success in an Anytime, Everywhere World* (NY: AMACOM, 2010), pp. 199–205.

23   Ann M. Morrison, "Check In, Check Out Hotel Review: CitizenM Hotel in Amsterdam," *The New York Times*, May 23, 2010 and www. citizenm.com/

24   www.megabus.com   and   http://articles.moneycentral.msn.com/ SavingandDebt/TravelForLess/MegabusThe1RoadTrip.aspx

25   Roman Friedrich, Michael Peterson, and Alex Koster, "The Rise of Generation C," *strategy + business*, Booz & Company, Issue 62, Spring 2011, p. 56.

26   Roman Friedrich, Michael Peterson, and Alex Koster, "The Rise of Generation C," *strategy + business*, Booz & Company, Issue 62, Spring 2011, p. 57.

27   Ian Tunnacliffe, "Google is Coming," *Airline Business*, June 2010, p. 69.

Chapter 5
Hurdles in Implementing New Enabling Technologies

Could the need for customer-centric information (relating, for example, to passenger needs and experience) and the up-and-coming technology (for example, digital and mobile) enable airlines to innovate their business models? Theoretically, yes. However, there are numerous hurdles that need to be addressed before most airlines can move further along in the innovation process. The enabling technology itself is already here, and improving each day, but there are many hurdles in the implementation of such technology. The overall challenge lies in the fact that the whole planning perspective appears to be unfocused and fragmented, as depicted in Figure 5.1. Specifically, the six major hurdles and their interrelationship are causing airlines to be unable to respond to its customers, let alone its competitors. In the long term, all airlines, but especially legacy airlines, will be forced to adopt the new-generation technology, partly, because it will be demanded by passengers and, partly, because it will be used by new airline entrants who are neither saddled with old systems nor conventional mindsets to develop hybrid business models. Moreover, airlines will feel pressure from a few high-technology companies that will use technology to introduce fundamentally different products and services in areas such as distribution. This chapter outlines the six major hurdles faced mostly by many legacy airlines in the implementation of technology that can help these airlines to transform their business models to meet not only their own needs (a reasonable and consistent return on investment and the ability to compete more effectively with low cost carriers), but also the needs of their passengers (not the average passenger or even different but conventional customer segments, but much more targeted segments, not to mention the possibility of each and every customer within a targeted segment).

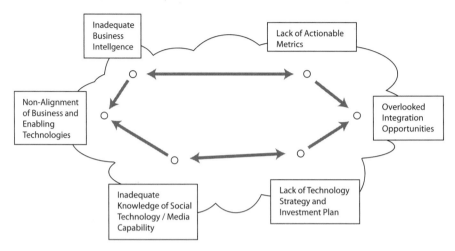

Figure 5.1   Unfocused and Fragmented Perspective

Why Are There Hurdles?

- First, there is still a debate within the C-suite on the view of what technology should be doing. Is it a cost center or revenue-enhancement capability center? Should it simply be a shop to go to ask for the development or acquisition of applications to facilitate the implementation of strategies selected by other groups, such as collection of baggage fees or the incorporation of a different way of the bidding process in crew scheduling? Or, should the technology group within the information department be a central place for a discussion on the economic and competitive viability of various business strategies followed by the development of the identified and prioritized business strategies and, later, followed by the utilization of appropriate metrics to evaluate the performance of the selected business strategy?
- Second, neither the whole airline nor individual functions have been able to develop a business case for deploying technology for the overall strategy or its individual components (such as business intelligence and actionable metrics). Again, take the case of distribution. While it may

be possible for an airline to compute the reduction in its distribution costs through the deployment of a particular technology, there are no reliable metrics for measuring the contribution of technology for an airline to access passengers, maintain control of its passengers, to gain greater intelligence about its passengers' buying behavior, or to predict the reaction of its passengers to the introduction of different value propositions by the airline and by its competitors.

- Third, many airline CEOs simply are reluctant to change their airline organizational structure that is necessary to deploy technology effectively (not just efficiently) to transform the business model. Some airlines have not selected the heads of their information departments who are both technology savvy and business savvy. Similarly, while members of the traditional "IT" department may excel in terms of technology know-how, some may lack the business knowledge that could help them leverage the technology more effectively. Next, while the "I" was part of "IT" (Information Technology) it should be separated from the "T" (making it Information *and* Technology) as information is the innovator (developer of competitive advantage) and technology is the enabler. Moreover, the CIO, if selected carefully from the viewpoint of a business mindset, should be sitting in the center stage of identifying potential strategies that are holistic and integrated. The CIO should have a 30,000-foot view of alternative strategies (and their integration) whereas heads of individual functions (marketing, sales, operations, and so on) may only have a 10,000-foot view of the proposed strategies. In light of the centrality of the CIO's position, especially in the current and expected structurally- and fundamentally-changing marketplace, airline Boards should consider making the CIO a member of the executive Board to improve the communication with and the understanding of the Board of the strategic challenges and opportunities on the one hand and to get their support on the other hand.

- In the old days labor was against the introduction of some technologies if their deployment led to a reduction in employment. Such a concern existed even within management. Within management, for example, the head of the traditional "IT" department would hardly be likely to support the outsourcing of all "IT" functions and activities and the ultimate closure of the "IT" department. The issue in more recent times is that the staffs in "IT" departments have tended to be much more risk averse to the point, in some cases, of actually impeding management in moving forward with specific business strategy initiatives. For example, staff members in an "IT" department could be so worried that a project might fail, that they may not even want to try to implement a new initiative. This risk averseness is the result of poor business intelligence, processes, and analytics, as well as metrics. Risk averseness is likely to be high in cases where the staff may not be really familiar with new-generation information and technologies, particularly consumer-related social technologies.

What Are the Hurdles?

Following are some detailed examples of the hurdles faced by airlines in the deployment of new-generation technology, that are summarized into six themes, as depicted in Figure 5.2.

Inadequate Business Intelligence

Business intelligence information is not being leveraged to its full potential within the airline industry relating to all stakeholders: customers, employees, alliances, vendors, authorities and regulators (shown in Figure 1.2). Even when limited versions are deployed, it is neither centrally based/centrally driven, nor user friendly. As with many other initiatives, business intelligence (applications and technologies) is based in silos and conducted on a piecemeal basis. For example, knowledge of passenger buying behavior or the costs of serving different types of customers or their profitability is very limited with many airlines. Even when this limited information is available in different functions (marketing, operations, and so forth), it is not shared. When the information is available in depth and breadth at an airline, it tends to be historic in nature as opposed to in real time. Even more importantly, business intelligence information is very rarely available with respect to the predictive perspective. Finally, the business intelligence capabilities are not in alignment with the metrics and the business and technology strategies, resulting in overlooked integration opportunities.

Proliferation of social media and technologies has broadened the channels of information and the depth and breadth of information. Information used to be available from traditional sources such as newspapers, radio, and television. It is now readily available in real time based on facts, opinions, thoughts, and so forth, and through channels such as Facebook, MySpace, YouTube, eNewsletters, blogs, Twitter communications, and so on. The end result is the availability of masses of information, in varying degrees of usefulness, coming from lots of sources, with a wide range of credibility. The user-generated information tends to be free but is usually unfiltered. The professionally-generated information is available at various levels of costs and for various situations and uses. Both types of information are entering the marketplace at an enormous speed. As Scott Klososky, an author and a business analyst, points out Twitter, for example, has now enabled ordinary citizens to become journalists, reporting on events in real time with news, pictures, and possibly even videos.[1] What is needed is the development of business intelligence from this explosion of user-generated and professionally-generated information that has been customized for the airline business, having been collected, verified, filtered, sorted, and aggregated with a broad spectrum of relevant social software tools (some online). Some examples of tools are Google Reader to keep up with blogs and news and Competitious to identify and organize information about competitors.[2]

Lack of Actionable Metrics

First, there is validity to the statement that most technology initiatives in the past have run over cost and have been behind schedule. However, this view can partially be attributed to the lack of balance and lack of metrics. Some projects were too big and too long and as such did not show results in acceptable timeframes. Mega projects either did not deliver or, if they did, they delivered too late. Moreover, the "IT" departments were held accountable for the task, rather than the result (metrics). Part of this criticism may also be related to the fact that managements tended to look to technologies and the "IT" department when economic conditions were challenging, and after the low-hanging fruit of cost-cutting had already been picked. Moreover, even when technology initiatives were developed to reduce costs or enhance revenues, managements expected an ROI within a very short period of time, months instead of years, and weeks instead of months.

Leaving aside the short timeframe for the expected ROI, part of the problem why a particular technology-based initiative might not have worked is not because of poor concept but because of poor execution and lack of metrics to measure the results. For example, airlines within an alliance may advertise "seamless" service but while one airline may be able to make a reservation on the flight of a partner, it may not be able to assign a seat or recognize the frequent flyer status of the passenger. Some computer systems still do not recognize frequent flyer status across alliances (or at least recognize accurately), and neither do employees. This feature of the gap between design and delivery defeats one important benefit of alliances. Most airlines have tried to make strides in rectifying this problem of lack of integration, in this case internally, by equipping their frontline employees with relevant data to assist them in executing their duties and measuring the results with agreed metrics.

The lack of appropriate metrics and analytics is now a critical problem in measuring the return on investments relating to social media and social technology. It is one thing to know how to leverage social media and technology. It is another thing to know how to put a value on its application. The concept of measurement of the value of social media and social technology is fuzzy since this whole area is relatively new. See the discussion in Chapter 2. If we assume that social media and social technology are here to stay and if we assume further the belief in the old saying that "what gets measured gets done," then the only viable option for an airline would be to gain insights from best practices.[3]

Overlooked Integration Opportunities

There is the issue of the new technology not working harmoniously with the legacy systems. Typically, airlines have numerous legacy systems in operation simultaneously. The legacy systems were developed for customized applications to work with static workflow processes and business rules. These existing systems do not even "link up" effectively and easily with each other, let alone allow users to work with dynamic workflows and business rules. Moreover, they are expensive to support (requiring, for example, maintenance of an aging programmer workforce) and have a long development cycle. The systems do not have a two-way relationship between products and processes. Current "IT" managements end up spending most of their time maintaining, on a day-to-day basis, the legacy systems and struggling with the interfaces that integrate the new systems (such as social media) with the legacy systems. Very little time, if any, is left to explore new technology-enabled products. Even worse, little time is devoted to exploring the impact of new technology companies that have heightened the expectations of 24/7 customers, as well as internal business and operational sponsors within airlines. However, while monumental, the task of changing over the legacy systems is achievable. In some airlines the "IT" group has not been willing to introduce new systems because they do not follow the corporate standard. For example, if an airline is working with a product of one vendor, it will not go with another vendor providing new services because employees have already been trained on the first product, licenses have been obtained, and so on. While an airline does need to have standards, it also needs to meet the changing business needs of the airline itself.

There is a lack of interconnectivity among various providers of service in the value chain. For example, consider "mobile travel concierges." Would it not be more user-friendly if the device was a virtual travel concierge and obtained the information not just on all flights available, but only in the next few hours, and from the same terminal, possibly even from the same concourse, so that the passenger does not have to go through the security process again? How about if the "mobile travel concierge" could also identify ground options for passengers booked on the last flight if their itineraries were feasible by a train, bus, or limo instead? Technology is available to integrate all transportation options. For example, a passenger should be able to get all options between Boston and New York, whether they be airlines, trains, or limousines from one source. The "mobile travel concierge" should not only be able to obtain information on available and viable options, but also be able to make reservations and even accept payments. There is one caveat, however. The air transport industry is highly interconnected through standards developed by global aviation bodies such as IATA and ICAO. Similar standards, that are a must, may not exist in other modes of transportation, thus making intermodal reservations more difficult.

Lack of Technology Strategy and Investment Plan

While mobile is clearly the direction where passengers are moving, mobile devices work on different platforms and have different capabilities and different functionalities. While Internet capability is available on many mobile devices, many of these devices are not user friendly. Examples include small screens, slow speeds, and poor reception. When most airlines developed mobile initiatives, some did not distinguish between the users of basic phones and smartphones. The customer experience that a passenger can receive from interacting with an airline's call center can also depend on the type of mobile device being used. For example, some questions may be better answered with the use of an image, such as a seat map or the location of a gate relative to a security area. The technology departments need to support a much more comprehensive array of channels and devices that are being used by passengers. Just as the technology systems adapted to the interaction with the Web, they did not integrate well with all mobile systems, especially not the higher end mobile systems. Moreover, the technology systems did not integrate well with the voice form of communications with mobile passengers, whether passengers were calling (speech) or texting. Even though technology is available, call centers were not able to respond to passengers through an analysis of their voice or visual interactions if smartphones were being used. This hurdle becomes more critical when one recognizes that even smartphones vary tremendously in their capabilities and functionalities. Given all of these differences within the mobile user segment, many airlines have not developed a business case for each of these sub-segments.

Airlines have been spending less than two percent of their revenue on information technology and communications. This amount may seem high for some carriers, those with a large revenue base and those who view technology as a cost rather than a center with the potential to enable the generation of revenue. The percentage amount is customarily based on an airline's previous year's budget rather than the need to develop and implement new business strategies. Moreover, the allocations are often based of history, say, for example, 60 percent of the budget on new projects and 40 percent on the maintenance of existing systems and based on the framework of a portfolio approach. This starting point can be a major stumbling block. The objective should be a decision on what is the strategy for technology that should go well beyond even the concept of multi-year funding based on themes as opposed to projects. For example, it could be to build capabilities for an airline to identify, prioritize, develop, and implement flawlessly its business strategies. The problem is not that airlines did not have a technology strategy in place. Rather, the issue lies in that the technology that they based their strategy upon kept changing rapidly. So, how could they have an appropriate investment plan given this challenge?

Many airlines do not have an information strategy with well-defined objectives and an enabling technology strategy. One consequence of this situation is that these airlines are not likely to have allocated their information and technology budgets within a highly-fragmented framework due to the continued functioning of the silo systems and silo thinking. For example, in most airlines, requests for technology support come from different functions on a project-by-project basis. Not only is this piecemeal approach inefficient, but often each project is not even supported with a business case due to the lack of actionable metrics (applicable both to operations and customer experience) as well as poor business intelligence, processes, and analytics. A group within the marketing department that is responsible for ancillary revenue may want technical support to implement charges for such services as changes in reservations and baggage fees. There is rarely a holistic and analytically based scrutiny that would include the negative impact on revenue from the potential diversion of passengers to airlines who do not charge such fees. In other cases, the operations department may come in with a request for technology to help with the scheduling of aircraft and crews on weekends as well as predicted weather issues. Again, the request is unlikely to be accompanied with a solid *holistic* business case and a piecemeal approach bears the risk that the assumed benefits in one department may, in fact, be compromised by initiatives in other departments.

Let us even assume that in both cases, the requests from the marketing department and the requests from the operations department are accompanied by standard businesses cases. In both cases, neither department is likely to know the implication of the request within the technology department. For example, the technology group may need to use different technology platforms to find quick fixes to meet the needs of these piecemeal projects, leading to an increase in complexity, and hence, costs. Each request represents a tactic (and even that on a piecemeal basis) as opposed to a long-term strategy within a holistic and integrated framework. Consequently, individual subgroups within the "IT" department end up working in isolation producing results that are sub-optimal (again, due to the existence of the silo system and silo thinking). Furthermore, each request not only strains the in-house technology infrastructure, but it also brings in layers of new technology components and vendors leading to a massive increase in complexity and costs. The problem becomes even more severe when some functions continue to want to work with their older systems (for example, reservation systems) even though new systems have become available in the marketplace. Most low cost carriers, however, used a different approach. When the Internet became widely available within the consumer community they switched all their customer-related end-to-end processes completely to the Internet format. As a result, they achieved substantial process cost savings without the need to work further with the old systems.

Finally, it is not just sufficient expenditures on technology systems; management have not tended to bring in top-notch employees with relevant skills, regardless of their age and lack of experience in the airline industry. They are the employees who truly understand the implications of trends, such as:

- online is now becoming mainline
- mobile traffic is increasing at exponential rates
- frequent travelers could easily have digital identities
- people are becoming more and more connected with no geographic boundaries or time limitations
- devices are becoming more and more connected with each other
- people are becoming more and more connected with devices
- devices are now becoming extremely application rich
- technology can not only enable an airline to become a solution provider after a problem occurs, but seek solutions by anticipating problems

Inadequate Knowledge of Social Technology/Media Capability

Consumer social technologies must be leveraged to take greater control of situations by being proactive rather than reactive. Remember the United Airlines guitar situation and also the JetBlue lengthy tarmac delay situation? Airlines are still struggling to understand the potential power of social media. At the present time, the focus appears to be the use of social media for crisis management. The situations can range from daily routine operations issues (such as lost baggage) to major crises (such as the volcanic ash issue, the upheaval in Egypt, or the earthquake/ tsunami in Japan). However, the marketing potential seems to be overlooked. The value in this area can range from customer relations to the sale of more tickets. As highlighted earlier, some airlines have begun to take advantage of these opportunities. However, the initiatives taken thus far are on a piecemeal and reactive basis. Moreover, one airline's initiatives tend to be based upon what another airline might be doing. The problem appears to be based partly on the lack of sufficient appreciation for social technologies/media and their strategic deployment relating to both passengers and employees. It is like the tale of the three blind men, all of them trying to figure out what an elephant is by touching a different part—leg, trunk, tusk—but only one part. Therefore, each has a dramatically different take on the object and there is no single let alone complete view of the animal.

The question is not whether social media and technologies are here to stay, but rather to figure out how to harness the power of social media and technologies.

Non-Alignment of Business and Enabling Technologies

There is the concern regarding spending money on technologies that may become outdated in a short period of time. Many airlines have tended to stick with old but proven technology systems rather than deploy something new—old reservation systems, for example, as opposed to systems with contemporary shopping and reservation engines. This hurdle ties in with the earlier point that technology departments tend to be risk averse. Moreover, there is a concern than some of the technology is a fad. Some airlines are skeptical of the role and impact of social media such as Facebook, Twitter, and YouTube. These skeptical managements do not see how social media can allow their airlines to become more transparent and more authentic, not to mention the more effective management of customer relations. Rather than looking at social media as a fad, they have not seen how they can be in control with the use of social technology by carefully leveraging the popular social media through carefully designed communication strategies. Even though social media is being used by passengers, competitors, and employees, some airlines have not embraced it to become transparent (and to more effectively manage complaints) and instead have resisted it with the thought that it is a fad. This hurdle has kept many airlines from integration opportunities such as connecting their websites with those of social media to introduce new products or to counter passenger complaints. In summary, the fad view of social technology and media has precluded some airlines from achieving the benefits of transparency and control. In the final analyses, it is social media that can enable an airline to better communicate and share information with passengers.

During difficult economic times, some airlines tend to implement strategies and tactics that unintentionally deteriorate customer experience instead of enhancing it to retain passengers and to build passenger loyalty during better times. These actions tend to be justified based on similar actions taken by competitors within the airline industry. For example, one airline cuts food on short- and medium-haul flights, many others follow. While it is recognized that mature industries need to cut costs, costs tend to be cut in two areas (those that affect the passenger experience and also technology budgets) whereas, in fact, enabling technology should be implemented to maintain and augment the passenger experience. This aspect is important given that passengers can now compare the service provided by one airline, not just with other airlines, but also with the service provided by other businesses. Yet, while transparency has become relatively easy in other businesses, it has become more difficult in the airline industry. When a passenger searches for a flight on the website of a typical airline, only selected seats are shown as available. It does not show that other seats may be available but they are either available at higher fares or made available only to passengers with a higher status in the frequent flyer mileage programs. Airlines should consider the value of gaining control by becoming more transparent.

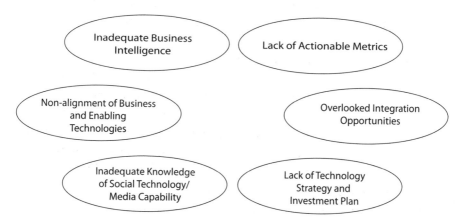

Figure 5.2   Summary of the Major Hurdles in Implementing New Enabling Technologies

These hurdles are slowing airlines from implementing capability-driven enabling technology strategies that can help them differentiate their services. After all, how can an airline appropriately market to the targeted passenger segments, value-adding collaborators, and real competitors if it:

- does not have a unified view of its customers?
- has management that is reluctant to understand, let alone implement, emerging consumer technologies (such as those discussed in previous chapters)?
- does not allocate sufficient amount for information and technology based on its size (number of aircraft, number of destinations, number of passengers), its complexity (diversity of networks and fleets and alliance membership), and its aging legacy systems that neither provide access to relevant and timely information, nor robust functionality?

Takeaways

There are serious hurdles that need to be addressed before most airlines can progress in terms of step-change in their business model innovation. The enabling technology is available, but the hurdles are holding airlines back from reaping the full benefits of the implementation of such technology. These hurdles can partially be attributed to the debate within the C-suite on the role of technology in the organization, the lack of alignment among the information, business, and technology strategies, as well as the overall hesitation to change the airline organizational structure to transform the business model. These hurdles, including inadequate business intelligence, lack of actionable metrics, overlooked integration opportunities, lack of technology strategy and investment plan, inadequate knowledge of social technology/media capability, and non-alignment of business and enabling technologies, can lead to the deployment of unfocused and fragmented initiatives within the organization.

Notes

1   Scott Klososky, *Manager's Guide to Social Media* (NY: McGraw-Hill, 2011), p. 103.
2   Scott Klososky, *Manager's Guide to Social Media* (NY: McGraw-Hill, 2011), pp. 106–8.
3   Scott Klososky, *Manager's Guide to Social Media* (NY: McGraw-Hill, 2011), pp. 145–57.

Chapter 6
Opportunities for Overcoming Hurdles

The previous chapter highlighted six hurdles impeding many airlines from leveraging the use of new-generation technology. Airlines can start to overcome these hurdles by recognizing that (1) technology that in the past was often referred to as "IT" must now be separated into two distinct parts: information *and* technology, and (2) there are four *interconnected* elements to consider (Figure 6.1) where changes must occur: business strategy, new points of integration, information strategy, and technology strategy. This chapter describes the transition from the current unfocused strategy (due to the hurdles discussed in the previous chapter) to the desired holistic and focused strategies called for in Chapter 7. Referring back to Figure 5.1, an airline needs to sharpen its focus in dealing with these hurdles, and set in place strategies to overcome them. An approach is needed that is sharp as a diamond, that is strong enough to overcome competitive resistance, and that shines in terms of customer satisfaction. This diamond is illustrated in Figure 6.1, in which the information strategy focal point is at the top of the diamond, while the technology strategy focal point is at the bottom of the diamond. On the left-hand side is the business strategy that is focused on continuous improvement, technology absorption and customer centricity, while the right-hand side are new points of integration and ways of creating value and profits.

Imagine the diamond shown in Figure 6.1 in this chapter set over top of the unfocused and fragmented state shown in Figure 5.1 in the last chapter. It would then become apparent how the current hurdles represent unfocused activities that need to be sharpened up, or reshaped, into the points of the diamond.

For example, business intelligence that is based on data that resides in silos and not integrated is the flip side of the lack of actionable metrics and lack of orientation around actionable data. Addressing those hurdles with a solid information-oriented focal point will sharpen an airline's capabilities and resolve those hurdles into a sharp point of competitive impact. Similarly, inadequate knowledge of social technology/media capability and lack of the technology strategy and investment plan represent ineffective activity that needs to be drawn to a solid technology focal point. Furthermore, on the left-hand side is the non-alignment of business and enabling technology that symbolizes wasted efforts and duplicative management thinking that, in turn, must be sharpened up into a business strategy of continuous improvement and customer centricity. Similarly, on the right-hand side are the overlooked integration opportunities which need to be focused into a finite number from which new points of integration can be chosen.

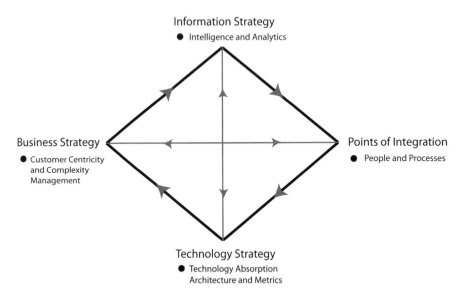

Figure 6.1   Focused and Holistic Planning Perspective

Information Strategy

It has been stated numerous times that airlines need to focus more on becoming customer centric than on product and operations centric. As stated in the Introduction, customer centricity means different things to different passengers. For some passengers, it means enhanced customer experience, and for others, greater control relating to purchase and delivery. From the viewpoint of airlines they need to know the degree of focus by segment. There might be the segment that represents 15 percent of the passenger base, 50 percent of the RPM base, but 90 percent of the profits. This segment needs not only superior products but also personalized services that are likely to be based on sophisticated information. This is the segment that calls for differentiation, at least based on information. The other segment, the much larger segment, may comprise mostly of price shoppers. Here the airline must be price- and service-competitive and there may not be a need to shift from commodity to differentiation. In both cases it is the availability of insightful and actionable information that can help an airline to become customer centric.

Moving down from a higher level use of information to develop strategies to a lower level to develop tactics, consider, for example, the use of mobile devices that range from simple phones to sophisticated smartphones. Just as there is variation in the sophistication of devices, there is an equal, if not more, variation relating to the technology know-how and technology usage. Think about the use of the mobile phone. The new generation uses the mobile phone in different ways from the older generation. Consequently, there is a need to obtain information on the expectations of different segments of passengers with respect to customer centricity. This need is closely related to the need to obtain information on customer behavior at different points in the travel cycle. How do different types of passenger behave while on the website of an airline? What kind of passenger is attracted to the website of an airline in the first place? As mentioned in Chapter 4, Web-user behavior needs to be analyzed not only in terms of final purchase behavior, but also the process in which the purchase was made. How is each individual passenger conducting his or her research? Is it always by price, or schedule, or combination, or something entirely different? If it is schedule, is it by time of departure, the duration of the trip, avoidance of certain connecting airports, avoidance of certain alliance partners, and so forth? How long did the passenger spend on the website? How many times did the traveler return to the website before actually making the purchase? Finally, what type of purchase was made? Was it a low-fare ticket or a refundable ticket? Did the information provided on the website lead to an up-sell and/or cross-sell? It is the acquisition of new-generation information that will lead to further innovations in the business strategies discussed above.

Following are a few examples of other companies outside the airline industry that have leveraged the information component.

- 7-Eleven Japan has leveraged information to empower all of its employees of its 12,000 stores with data in real time such as product information, recent sales, and even weather information. Store staff utilize this data to make decisions regarding orders for their store, and receive feedback on the results of their business decisions, which in turn sharpens inventory- and revenue-management decisions and, ultimately, yields highly efficient operations as well as product innovation.[1]
- UPS has leveraged information to better understand the profitability of customers and packages at the individual level. This initiative, in turn, has yielded more informed decisions in terms of both pricing and routing. Specifically, UPS's drivers are equipped with delivery information acquisition devices that not only capture the customer's signature, but also contain data on each delivery that feeds into its database.[2]
- Isle of Capri Casinos, which features entertainment, gaming, and lodging at over a dozen properties in the US, faces a unique challenge due to the fact that the company has properties in six different states (Colorado, Florida, Iowa, Louisiana, Mississippi, and Missouri) and therefore a diverse clientele. This challenge involves the delicate balance of customer segmentation while maintaining the brand. In order to achieve one of its goals, to build loyalty through

enhanced and timely marketing efforts, the company invested in a data warehouse and CRM tool (through Teradata) and a BI platform (through IBM Cognos). While the company initially hoped to augment its direct mail initiative (that was piecemeal based on different properties) and obtain a single view of the business, new analytics capabilities enabled the company to also consider other areas such as hotel and slot machine data, ultimately yielding much deeper insights into customer behavior.[3]

Understanding technology as an enabler for the acquisition of new-generation information (for example, Web traffic and Web behavior) can lead to the development a different perspective for technology. Although marketing has traditionally been responsible for determining the changing needs of passengers, it is the technology group that is better equipped to use new-generation technology to get new-generation information to gain insights into not only the changing needs of various segments in the marketplace, but also changes required in the processes to design and deliver the relevant products and services. For example, travelers' stress levels are at an all-time high due to the pressures of the always connected, 24/7 world. Most travelers are severely time-constrained. Passengers are concerned with the length of time it will take to clear security, the length of time it will take to deplane, missing their connections, and lost baggage. The technology group can use new-generation technology to get real-time information on why passengers display certain behaviors in the travel process. For example, why do they choose the flights that they do? This concept was touched upon in Chapter 2, but this discussion focuses on a different aspect. The question raised in Chapter 2 involved whether a passenger chose a particular flight because that is what he really wanted, or because that was the only non-stop choice available, the only flight in the timeframe needed, and so on. The question being raised in this discussion is why did the passenger choose the 6 AM flight when there was a perfectly good flight at 9 AM that would still have allowed the passenger to make his connection? One likely explanation for the passenger choosing to fly three hours earlier is to have more buffer time in order to ensure he makes his connection.

The technology group can help marketing examine these types of behaviors, gather information, and maintain data in an attempt to reduce travelers' stress and improve the passenger experience. There is a real opportunity for many airlines (and in some case, airports) to leverage technology to help remedy this situation and help alleviate some of the stress that burdens travelers. For example, passengers with tight connections could be seated at the front of the plane, regardless of their status in the frequent flyer program. Carts could be stationed at the arrival gates of passengers with tight connections (for passengers on delayed flights), ready to drive them to the gates of their next flights, as mentioned in Chapter 3. Even prior to arriving at the airport, passengers could receive text alerts with approximate current wait times for the security lines for the terminal in which they will be traveling. Finally, as addressed in Chapter 3, passengers could receive texts regarding the arrival of baggage on a different flight, thus avoiding not only an unnecessary wait at the carousel, but also at the lost-baggage office.

Information and new-generation technology can also be leveraged in the area of security. The air transport industry and governments have been collaborating to create a more secure air transport system since September 11, 2001. However, there are still challenges, including passengers having to wait in long lines at security screening areas as well as some security incidents that have occurred in the last decade. As SITA points out, accurate risk assessments are needed for each traveler, an endeavor that can be reinforced by intelligent information and technologies. Such an intelligent-led approach involves a shift from random inspection to selective targeting (based upon the aforementioned risk-adjusted assessment). Such intelligent-led solutions (based on information and enabled by technology) should be more efficient and effective, while also being less costly.[4]

Suppose a carrier has a hub at Frankfurt Airport. Suppose that of the many flights arriving at Frankfurt Airport three have been delayed. There are four passengers on these three flights connecting to the carrier's London flight that is the last flight of the day of this carrier as well as any other carrier. Unfortunately, there are only two available seats, so two passengers must stay overnight. Two can go tonight, and two must wait until tomorrow morning but even on that flight there are only two seats. Currently, the carrier has typical rules that reservations can use to determine which two passengers would be chosen. These rules tend to be based on such criteria as frequent flier program tier status, the price of the ticket, and the point of origin. What could be used to improve customer satisfaction ratings would be a better rules-based engine to include more real-time factors and input from each of the four passengers. This case, developed by Teradata (BSI: Teradata Webisode: The Case of the Mis-Connecting Passengers), is presented in much greater detail in the Appendix, as the decision keeps changing with the availability of new information on each of the four passengers.

As this case illustrates, if the first component of information strategy is intelligence, then the next component is analytics. The key to customer analytics is arriving at the individual behavior level. Most airlines tend to consider averages, rather than individuals. When a particular customer calls in, or logs on, the airline should already have a good idea of the maximum amount that this particular individual is willing to pay. Furthermore, the carrier should have even a clearer picture regarding the individual, such as that he or she always takes the lowest fare, but then always accepts the upsell of an upgrade to business class. Therefore, there is actually a potential to make money on the individual. Perhaps the individual always only desires non-stop choices, in which case, the airline should not even try to sell him on the cheapest flight if it involves one or more layovers. Enabling technology can help identify these trends at the individual level in terms of both the core price of the ticket and ancillary revenue (upsell and cross-sell). Consequently, technology can enable an airline to implement choice-based revenue management practices.

Technology is now available to perform analytics even on unstructured data. Whereas structured data is normally organized by variables such as socio-economic characteristics, advances in the analyses of unstructured data (for example, text contained in emails and SMS) can provide significant insights into passenger behavior. Similarly, passenger calls coming into call centers are usually recorded and technology is now available to analyze these calls to obtain deeper knowledge of passengers' concerns (based on not just words but also voice tone) and to provide solutions to passengers' problems, not to mention an opportunity for up- and cross-sells. Another way of describing the differences between structured and unstructured data is that structured data comes from standard surveys whereas unstructured comes from comments made in the surveys or via emails, and so forth.

Points of Integration

As it is now commonly acknowledged, money can be made at the point of integration. Therefore, management needs to determine where the next points of integration lie. The starting point for this exercise is to the integration of information itself. In most cases, information coming from different channels of communications within an airline (call centers, Web, mobile devices, and social media) is not integrated. Even within an information-and-technology savvy airline, these channels represent silos controlled by the heads of different functions, resulting in a view of the passenger that is not unified (a hurdle identified in Chapter 5). This lack of a unified view makes it difficult to engage with a passenger effectively and to build a trustworthy relationship. One way of crossing silos is for an airline to have an event-driven view. For example, if a flight is late, all departments are involved, not just those directly affected. Another means of crossing silos is to capture the customer dialogue. For example, what was the reaction of a passenger when the airline provided more information? Key words, attitudes, and actions could be gained for the purpose of predictive analytics. Again, this analysis can be performed at the micro level rather than the traditional macro level. Technology is available to model future business scenarios that produce robust results with greater agility and greater transparency. However, again, there is a need to change the culture and move away from the traditional "IT" department. For example, the group must not only synthesize the data (from the viewpoint of consistency, accuracy, and so forth), but also to answer the right question. This is one example that illustrates technology strategy supporting information and business strategies.

The business strategy can therefore be formed around the opportunity to solve passengers' problems by integrating the fragmented technology and information systems and making an airline much more customer centric. The problems arise not only during irregular operations, but also during normal operations. Consider two examples during regular operations and one during irregular operations.

- An airline could plan to have access to hotel and car rental reservations systems for a short period of time while interacting with a passenger to accommodate changes that may need to be made at the time of booking. As mentioned in Chapter 1, if a traveler contacts an airline to travel to a destination for a weekend and books a round trip flight leaving Friday and returning Sunday, but then contacts the hotel only to find out that there is no room available for that Friday night, the traveler must then start the process over again and most likely incur a change fee for the airline ticket. Technology can be developed for an airline reservations agent to have access to the reservation systems of other travel-related businesses (hotels and car rentals) and other modes of transportation (trains, buses, and limousines) to provide comprehensive solutions-oriented itineraries to travelers, at various price-service options. At the very least the airline reservations agent could stay in the loop until the entire itinerary is complete. If the issue becomes one of higher costs of the agent's time, passengers could be offered opportunities to pay for such services.
- Frequent travelers have experienced problems that occur even when the airline statistics show that there are no problems. As mentioned in Chapter 4, while most airlines keep statistics on their delayed flights, many do not keep statistics on the number of passengers who missed their connections by cause. For example, an airline would know the number of misconnects during the day. However, what is not known is the cause for the missed connection. The inbound flight arrives on time. The passenger has a 30-minute connection time. However, the inbound flight on a regional airline delays the deplaning process until the bags checked in at the gate have been off-loaded and placed at the steps of the aircraft. If this process takes ten minutes and the outbound connecting flight is from a different terminal the

passenger misses the connection. There is an opportunity to change the process. Why not allow the passengers with hand-held baggage only to deplane while passengers with baggage checked in at the gate wait for their baggage to be brought to the steps of the aircraft?

- A passenger makes a booking through a global travel agency for a three-segment transatlantic trip involving two carriers within the same alliance. The first segment is cancelled; thus begins a cascade of problems due to the existence of fragmented processes. The passenger talks with the agent for the carrier whose flight was cancelled and is informed that the passenger must talk directly with the first carrier. The passenger, having waited a long time on the phone, does get in touch with the carrier scheduled to fly the first segment and is informed that he must call the carrier who issued the electronic ticket. After another lengthy delay, the passenger is told by the second carrier that the whole process must be handled by the travel agent who made the booking in the first place, or by the carrier flying the first segment. At this point the passenger is back to square one and the whole trip is not only delayed by 24 hours but the passenger ends up canceling the entire trip due to the fact that the passenger would miss the scheduled meetings that had taken weeks to arrange.

Leaving aside the lack of integration within an organization, there is also an opportunity to improve integration with external organizations to drive value. If an inbound flight arrives late and the last connecting flight has departed, an airline's reservation system could easily be connected to the reservation systems of alternative modes of transportation. Instead of simply rebooking the passenger on the first flight out the next morning, the passenger could be offered alternatives such as a reservation on a bus, a train, or with a taxi or a limousine. In many cases the passenger is not even provided with information on available possible transportation, let alone be given the offer to actually make reservations. As previously mentioned, an example of technology-enabled external integration is for an agent in the reservation department of an airline to actually have access to the reservation system of a hotel or car rental agency for the duration of time while the passenger is making the reservation. Currently, passengers are transferred to the websites or reservation departments of travel partners. However, this process requires the passenger to provide the same information again such as the credit card information for payment, but much more important, a conflict with the reservation just made for air travel. For example, a passenger makes an airline reservation. Then, while making the reservation for the hotel is informed that had the passenger been willing to arrive the day before or the day after the price for the hotel room would have been much lower. At this point, the passenger can accept the higher-priced room, go to a different hotel, or go back to the airline and change the airline reservation and most likely pay the change fee. Also, if an airline knew about some flexibility of a passenger it could offer him to fly one day earlier (freeing up a very popular travel day) rather than have the passenger stay in a hotel (to be paid by the airline). Technology is available for each member in the travel chain to have access to the reservation of the other for the duration of the reservation process lowering the total cost for the passenger and improving the total shopping experience.

Assume that an airline offers all of its high-tier passengers a free travel concierge membership (costing, say, US$50 a year) plus an online real-time concierge service by the airline that tracks the passenger while en route. As this travel concierge is likely to know all of the traveler's plans before and after the flight, his whereabouts and schedules, the airline can react in all cases of disruptions in a very personal and proactive manner. Moreover, since the travel concierge will undoubtedly constantly communicate with a traveler about delays, an airline could use this channel to not only provide information about delays, but also offer remedies targeted at the individual requirements. The actions could even mean rebooking the passenger on a competitor if this is the only chance to get the passenger to his destination on time.

As discussed in the previous chapter, back in the old days, it was thought that a business could only achieve one or two of the three aspects of production, namely, producing a product better, cheaper, or faster, but rarely all three. However, through new-generation technology and changes in current processes (end-to-end process management aligned with business strategy), it may be possible to work toward achieving all three attributes. Consider the example of Tetra Pak, a Swedish global food packaging company. The company was struggling with its processes, which were fragmented as the focus was on trying to keep up with demand rather than taking the time and effort to develop streamlined processes. There was a lack of discipline, a lack of alignment between processes and strategy, and a lack of vision in terms of the entire end-to-end process. Over the years, several attempts were made to rectify the problem, and while some progress was made, they ultimately did not provide a lasting solution. However, in 2003, a chief process officer was named who spent significant time bringing order to the myriad of previous initiatives. One of the key changes included aligning the processes with the company's main business strategies: (1) growing the core, (2) focusing on cost-driven innovation, and (3) driving operations and performance. Furthermore, the entire organization received education in terms of processes, starting at the top level, which consisted of 250 senior executives. Moreover, the company worked with the Swiss graduate business school IMD to offer a "process academy" in which executives received small-group training in process design in an effort to gain understanding and acceptance. The success of the process overhaul was attributed to a number of factors, including full commitment of the CEO, who, in turn, announced the process owners and their role. Other key factors include avoiding the desire to change everything at once, but rather to focus on a few core processes that are in alignment with the business strategies, as well as the role of executives in terms of getting employees engaged (with respect to process design) by appealing to them both intellectually and emotionally.[5] For more information on this subject, refer to the book, *Faster Cheaper Better* by Michael Hammer and Lisa W. Hershman.

Technology Strategy

Technology strategy, if developed and funded strategically, can build an airline's capabilities not just to enable, but to drive its business strategies. Figure 6.2 starts with that assumption in the left-hand box and then shows how the technology strategy can be linked to the business strategy. It is the technology-enabled capability that can allow an airline to identify the optimal target segments. It is also the technology-enabled capability that can show how an airline can develop and implement the identified business strategy relating to the required market intelligence, the approach, the processes, and the metrics required to measure the success of the business strategy.

Consider, for example, how technology can enable an airline to identify an optimal strategy for the distribution of its products and services, not just from the viewpoint of savings in costs, but also from the viewpoint of maintaining control of the customer, enhancing loyalty, and introducing a product differentiation. While the traditional channels were airline call centers and brick-and-mortar travel agents, now the channels include Internet websites, online travel agents, and some airlines even have mobile applications available for travelers. Passengers used to be checked in by agents. Now it is kiosks and mobile phones.

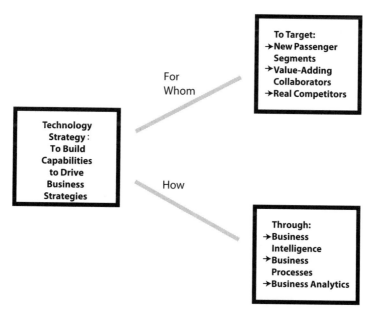

Figure 6.2   **Technology Strategies and their Enablement of Business Strategies**

Passengers used paper tickets. Now it is electronic tickets. In-flight entertainment consisted of magazines and movies shown on a few screens at various locations in the cabin. Now it is on-demand movies, live television, and access to the Internet. Even on the operational side, pilots carried their heavy manuals in their flights bags. Now it is computers in their flight bags. So, while there has been an enormous increase in the capability-building feature of technology, many airlines have struggled to keep pace with the expenditures on technology, to develop a comprehensive technology strategy to enable business strategies, and to align the technology strategy with the business strategy.

Some airlines are developing a greater technology know-how to leverage technology on a consistent basis to augment their operational and commercial performance. They are recognizing that technology can be at the core of the decision-making process and data that can be derived from transactions and processes and that can be leveraged to design and deliver the product more efficiently and effectively. These airlines are also gaining insights from other technology-savvy businesses. Following are a few examples of other companies outside the airline industry that have capitalized on technology know-how:

- Volvo once was plagued with a legacy system challenge. Specifically, Volvo desired to gain a better understanding of the mechanical performance of the company's vehicles under actual driving conditions. However, due to the presence of a legacy system, the company could not combine diagnostic readout data with design and warranty information housed in the existing data systems. To overcome this challenge, Volvo consolidated product design, warranty and diagnostic readout data into a data warehouse. As a result, several divisions (product design, manufacturing, quality assurance, and warranty administration) all have access to the same product data, fostering the company-wide shift to a fact-based decision-making framework.[6] Volvo overcame its silo thinking and, in turn, its legacy system challenge, and ultimately achieved integration to drive value. While this initiative seems like an obvious solution, many airlines simply do not take a similar approach. Some do not spend money on integration. Some continue building fiefdoms that, in turn, foster silo thinking. Many continue to think that their situation is unique (compared to other businesses), and therefore miss the value of such initiatives. On the other hand, other travel service providers, such as hotels, are beginning to leverage technology at a quicker pace relative to airlines.

- Mercedes-Benz recognizes that its consumers are using mobile devices to manage their "on the go" lifestyles, and has therefore implemented an enhanced version of its mbrace Mobile Application accordingly. This latest version connects the car's navigation system with the driver's iPhone, thus enabling the driver to access Mercedes' Concierge service, even when the driver is not in the car. For example, one can discuss restaurant recommendations with the service over his iPhone while out and about, and then the Mercedes-Benz Concierge can send the information to both the driver's iPhone and his car's navigation system. Moreover, the flexible architecture enables Mercedes to offer new services as mobile technology evolves, thus allowing the company to be a leader in terms of innovation and connectivity.[7] Mercedes-Benz is certainly a good example of a company that understands, plans for, and implements internal-external systems connectivity, all in a simple and seamless manner to better serve its objective, serving its customers' needs. The company's holistic framework, in this case, its flexible architecture, allows it to augment its offering and therefore maintain its relevancy as mobile technology evolves.

- Consider an example from the financial community. ING DIRECT offers core banking products directly via Internet, mail, or phone to customers of country bank organizations in Asia, Australia, Europe, and North America. Since each ING DIRECT customer is a customer of a specific country bank, the company's unique framework does not require a shared data platform across the countries. Rather, its business model involves a global brand that consists of simple high-value banking products for consumers residing in various countries. The company has established a single platform that contains one standard suite of bank products, business processes, infrastructure, and systems that may be implemented at each country bank.[8]

Even within the broader framework of the airline industry there are companies that are leveraging technology to offer value-added services for the airline industry, thus dramatically changing the traditional travel landscape. Dohop, an Icelandic technology company, originally proposed to connect all the low cost carriers' offerings for a passenger through a search engine. However, the company, who was already equipped with legacy carrier data, decided to provide both. In the original concept, for example, if a passenger stated that she wished to go from A to B, Dohop aimed to be able to figure out how the passenger could do so by only utilizing low cost carriers. The traveler would just input her travel information and then the search engine would provide the price-service options. The passenger can be connected to the website of the relevant airlines. The service was launched partially due to the fact that one of the founders lived in France and invested a lot of resources trying to get to Iceland.[9] Another example would be the emergence of external companies that will actually move baggage for passengers for a stated fee, much like FedEx currently does for packages. Such offerings by external companies can provide airlines with some insights to move forward in terms of their technology strategies.

Beyond the core product airlines can use the technology to offer a wide array of customized services ranging from the use of the Internet to entertainment (including gambling) to food to shopping (with various delivery options). New-generation technology can not only enable an airline to offer customized services on-board, but also enable the airline to capture vital information on shopping and purchase behavior for fine-tuning business strategies for the *captive* audience. Technology is advancing at such a high pace that one partner in an alliance should be able to provide customized on-board services on its own flights to passengers booked by another partner or a travel agent.

At one end of the spectrum could be airlines that want to focus on the value of getting the basics right, a topic highlighted in Chapter 4. For them, one solution could be to leave complex issues, such as technology, to third parties, such as Google, Kayak, and TripIt. For these airlines such an approach could be better than deploying technology as a "bandaid" solution to remedy fragmented problems rather than developing and funding a comprehensive technology strategy as an enabler for its information and business strategies. At the other end of the spectrum may be airlines that may want to develop a common platform operating within a digital framework (so as to be able to accommodate the new-generation technology). Take, for example, Lufthansa, an airline with multiple subsidiaries. While each subsidiary is operating with its own brand, the technology foundations can be used to manage all subsidiaries from a single platform, similar to the experience of ING DIRECT. Recently, some airlines have been exploring the deployment of new-generation technology to bring about significant changes in the distribution channel, as discussed earlier in this chapter.

At the other end of the spectrum could be airlines that may want to focus on the latest enabling technology. For example, Web check-in has provided significant advantages for passengers traveling in domestic and some regional markets. Technology is now enabling airlines to offer mobile check-in in these markets. The next frontier for hassle-free travel is to simplify the check-in process for international flights where passengers are still required to check in manually (progress is already being made toward this effort in Europe, but it is still needed in many other areas including the US). This is one place where information, technology, and innovation must come together to bring about a significant improvement in customer experience. Technology is available to (a) integrate and standardize information relating to government requirements for passenger data and documentation, and (b) facilitate collaboration between government departments, airports, and airlines. Just imagine the type of services that a major legacy carrier (with offices around the globe) could provide—meet and greet people, acquisitions of visas, and so forth—all for various levels of fees (ancillary revenue). Flydubai, a division of Emirates, is reported to already have made arrangements with the Dubai Naturalization and Residency Department to acquire visas by email or mobile phones to speed immigration process for passengers coming from countries that require visas.[10] Of course, there would be added complexity and higher costs. However, there would also be higher revenues, a step increase in loyalty, and a de-commoditization of airline service.

The first consideration is the need to implement a capability-enhancing technology strategy. The next concern is how to fund such strategies. Clear objectives and criteria need to be set at the executive level in terms of the technology strategy and its role in implementing the information and business strategies, as discussed above. The funding for technology needs to be separated, the part to maintain existing systems and applications, the part to create a flexible technology absorption architecture, and the part necessary to remain committed to the absorption of the architecture. For example, with respect to social media, the architecture would involve the deployment of the systems, applications, and people.

Second, clear metrics should be established for evaluating the performance of technology and its enablement of different components of the information and business strategies. Returning to social media, in addition to focusing on whether it will have a return, airlines should be thinking about how to get the return. Furthermore, consider, for example, the role of technology in helping an airline convert its legacy transaction-based reservation system to a more contemporary shopping, booking, and profiling system. The current transaction-based legacy system basically contains data on a passenger's name, record locator, and the ticket price. Since an airline would typically already own such a system, the only expenditures would be related to the maintenance of such a system and payments to external support providers, such as GDSs. If an airline were to acquire and implement a more contemporary system that could provide a better shopping experience, it could result in an increase in the online booking share for the airline. Furthermore, if the contemporary system were also to contain a profiling system that reflects some unique attributes relating to a passenger, the online booking share would increase even more. This, however, also requires product or service initiatives responding to known profiles. These initiatives should either help increase the passenger's loyalty beyond that of a typical frequent flyer or they should help "up selling." Known profiles without the capabilities to react accordingly are not very useful and explain why some airlines were frustrated after having invested in CRM systems. Examples of profiling attributes and reactions can include:

- as mentioned in Chapter 4, a passenger who typically flies in business class, but on this trip is flying with her family in coach could be offered an upgrade at favorable cost in return to extending her FFP membership
- a passenger who regularly takes this flight could be offered some sort of a "go show" privilege (no need to book, just come and fly)
- a passenger whose baggage has been lost several times in his last few trips could be offered a more generous carry-on allowance
- a passenger whose luggage did not make the flight due to a tight connection could be warned, if booking another tight connection
- a passenger with top-tier status who experienced one delayed flight and one cancelled flight in his last five trips should be met by the airline's station manager to discuss his recent mishaps (also a good learning experience for the station manager)

Incorporating such a contemporary shopping, booking, and profiling system would enable an airline to get much closer to becoming a solution provider and a value integrator, leading to an increase in passenger loyalty. Consequently, the costs of acquisition, implementation, and maintenance of such a system should be evaluated relative to the benefits. Therefore, as mentioned in the previous chapter, technology initiatives need to be funded based on the contribution of technology to enable the fulfillment of business objectives rather than on historical budgets. If the total cost of the shopping, booking, and profiling system is an order of magnitude less than the value it would bring to the top line, then the technology budget should be relatively easy to determine.

New-generation technology can provide new ways of accessing and connecting with old and new customers. Consequently, technology must be viewed in light of customer access and customer-centric product delivery and not simply as a cost center. And it is this understanding of the linkage between the execution of the business strategy and the enabling capability of technology that will eliminate the hurdles highlighted in the previous chapter. Yet, within the airline industry, some carriers have "outsourced" their IT operations to an internal subsidiary working as a profit center. This scenario can lead to a "P and L mentality," that may not foster investments in new technologies or processes, unless specific departments submit requests. Insights provided by other businesses will go a long way in persuading airline managements to develop a different mindset relating to technology, namely, a profit rather than a cost center. The Volvo example illustrates the potential for airlines to overcome their hurdles.

Returning to the challenge of the lack of metrics to anticipate the actual ROI of investments in technology, although difficult, such metrics can be developed, ironically, with the use of the new-generation technology itself. Take, for example, Google whose business approach appears to be metrics-based. Chairman and CEO Eric Schmidt illustrates this quality in stating that the company knows that if one spends X dollars on advertising, then one will yield Y dollars in revenue.[11] This approach, however, has not necessarily been the case in traditional media that has been driven more by guesses and emotions rather than metrics. Consider the case of the Super Bowl. In the traditional advertising arena, a company would spend several millions of dollars for a 30-second advertisement spot, not knowing the expected returns. Google's approach involves facts, beta testing, and mathematical logic. The result is a more rational and transparent advertising business model. Google also has implemented strategic business intelligence and analytics, as Google has assisted advertisers in targeting consumers by personal preferences in terms of activities, products, places, and so forth in addition to the traditional targets such as age, income, sex, and zip code.[12] Consequently, there are significant insights that can be gained by the airline industry from the experience of other businesses to develop and deploy analytically-based metrics to evaluate technology, information, and business strategies. One way to use this knowledge on a passenger's profile would be the ability to "program" the individual IFE screens that have become commonplace on intercontinental flights to show "targeted" advertisements, thus generating much higher revenues from the advertisers. Consequently there is a need not only to measure against known metrics, but also to consider new metrics.

From a broader perspective, rather than individual business units approaching the "IT" group, the information and technology group should take the leadership to work with all areas of the airline to review the overall technology strategy, highlight capabilities, determine how technology can be better leveraged to support the capabilities, and ultimately, the overall business strategy. It is through such a proactive leadership that the technology group can link technology strategy with business strategy and manage demand between functional business requirements and technology capabilities and capacities. This demand management aspect is particularly relevant in dealing with functional barriers and alliance partners. The irony is that management can, in fact, deploy technology itself to break down the silo-related barriers and bring about common standards among strategic alliance partners. In the case of the latter point, think about paradigm-shifting technology that allows each strategic alliance partner to maintain its own technology system, but at the same time uses technology itself to coordinate the different systems, when and as needed, for example, the transfer of data, regardless of the format, the source, or the location of the data. However, such a radical scenario would call for a major shift in culture—the willingness to feed the "virtual" technology integrator, the willingness to move away from conventional thinking, and the willingness to invest in the information infrastructure.

Business Strategy

The starting point of the business strategy is to determine, from the customers' points of view, their contemporary needs (such as hassle-free travel). This requirement involves a much greater focus on customer centricity that, in turn, links back to finding the next points of integration that create value (hence the two arrows between business strategy and points of integration shown in Figure 6.1). In the past, examples of points of integration included network, fleet, and schedules, as well as the development of alliances. Two current examples of points of integration are changes to the distribution strategy and the introduction of new seats in the economy class in long-haul flights.

Consider first the case of an airline's business strategy to bring about a significant change in the way an airline merchandizes its products and services. The first objective could be, as shown in Figure 6.3, to increase an airline's profit margin partly by decreasing its distribution costs and partly by increasing its ancillary revenues. Costs can also be decreased by reducing the capital employed. Typically, when passengers are on the website of an online travel agency they see schedules and fares of different airlines displayed via the systems used by GDSs. When a passenger decides on a seat on a particular flight the agency makes a booking on the selected airline through the GDS. The airline pays a commission to the agency and the GDS. The GDS then also pays a part of its commission (from the airline) to the agency. If the airline were to use technology to enable the agency to connect directly to the airline's reservation system (bypassing the GDS), there would be a reduction in the distribution costs. In addition, if the airline were to merchandize its products and services directly, it could increase its ancillary revenue through the development of a broadened portfolio of products and services, enabling up-selling and cross-selling.

The second objective could be to gain back the control of the customer that airlines have been losing over the years, as mentioned in Chapter 4. The increase in control could be achieved by engaging directly with passengers to determine their needs and to keep track of their buying behavior. Moreover, through interactive engagement and knowledge of passenger profiles, an airline could begin to merchandize personalized products and services, encompassing a process of bundling, unbundling, and re-bundling. This objective would also move an airline from offering a commodity product to one offering distinctive products. This business strategy has a potential to bring back some pricing power and build brand equity. Consequently, the objective may be to choose more effectively an airline's distribution channels and to provide passengers the elements of the product they want based on their willingness and ability to pay.

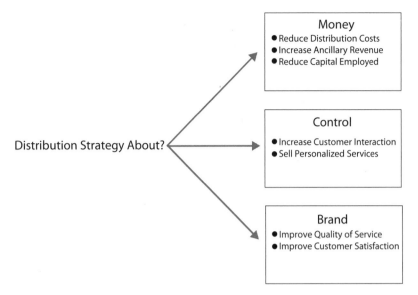

Figure 6.3   Example of a Technology-Enabled Information and Business Strategy

The third objective could be to enhance the brand by improving the quality of service and enhancing customer satisfaction. One means of achieving these goals in the pre-travel phase is through the development of more user-friendly websites. Another way would be to augment the airport experience through leveraging mobile technology to provide location-based services. It also enables an airline to select the distribution channels with which they conduct business, which, in turn, distinguishes the airline brand.

The successful implementation of a business strategy requires not only the use of new-generation technology, but also the direct involvement of the technology group right from the beginning. The technology group can identify the new relevant technology, its linkage with existing technical systems and the development of metrics as well as the data to evaluate the viability and sustainability of such a business strategy. Consequently, personnel in the technology group, assuming that they are also business savvy, can add significant value to the development of a business strategy and not just its implementation.

   Another example of a change in business strategy could be to target a different segment. In intercontinental markets most airlines have and are focusing on the needs of passengers traveling in premium class. This is a perfectly legitimate decision in light of the presumed profit margin of this segment. On the other hand, an airline could also consider distinguishing itself by targeting the needs of economy-class passengers traveling in ultra-long-haul markets. In this case the technology required might relate to the design of a seat and the configuration of the cabin. The technology group can help in the holistic evaluation of such a business strategy, all the way from segmentation to profit margins.

   Consider the innovative and customer-centric initiative by Air New Zealand to develop a lie-flat seat in its economy-class cabin by converting a set of three normal economy-class seats into a relatively lie-flat seat. See Figure 6.4. The three leg-rests come up in a horizontal position as shown in the photograph. An underlay is then placed over the seat-pan and the leg-rests. The passenger is then provided with a sheet, a blanket, and a pillow. This initiative is clearly an example of a technology-enabled business strategy.

   Once technology has been used to develop a new seat, the next challenge is how to sell it. There are passengers who simply may want to sleep. There are parents who want to use the seat as a couch for small children to play. Then there are those who just wish to have more space as well as privacy. Next, how can technology be leveraged to obtain passenger feedback, in real time, with respect to the experience and the price paid? The challenge is to view the new product from a holistic point of view—the cost to the airline, the value assigned by the passenger, and the willingness and ability of the passenger to pay.

   Outside the airline industry, MinuteClinic would be a good example of a business strategy relating to intelligent segmentation and a point of integration within the healthcare industry. MinuteClinic (which was originally called "QuickMedx" and is currently associated with CVS, a US drug store chain) was established after a man sat for several hours in a traditional health care center waiting for a simple medical test for his son. He knew there had to be a quicker, more convenient method to take care of routine medical problems. Thus, MinuteClinic was founded upon the following principles:

- easily accessible (no appointment needed)
- affordable (a list of the service offerings complete with a range of the costs associated with each offering is provided online)
- convenient (open 7 days a week)[13]

Figure 6.4   **Air New Zealand's Lie-Flat Seat in the Economy Cabin (printed with permission)**

The clinic does not try to be everything to everyone, but rather is limited to performing a handful of procedures and tests by trained family nurse practitioners and physician assistants. MinuteClinic is a good example of a company that possesses a well-defined business strategy: offering a few routine medical procedures rather than trying to be everything to everyone. The point of integration would relate to the integration of a limited number of, but carefully selected, doctors (located elsewhere) who are paired with a large number of nurse practitioners with strategically selected backgrounds.

What if a small niche airline tried a similar scenario? What if such an airline were to cut back to the top 19 markets, eliminate reservations, and offer shuttle service in those top 19 markets? This would drastically reduce complexity and also costs. Currently, many airlines are overwhelmed by self-imposed complexity, some of which could be reduced by technology, and some of which could be reduced by doing what others, such as MinuteClinic have done, focusing on the top medical issues, or in this case, markets, and not trying to be everything to everyone. Next, what if another small niche airline were to develop a strategy to actually function more like a high-speed rail system? Such a scenario could address several of the hurdles that the airlines currently face. First, there would be no seating assignments. Passengers could purchase tickets ahead, at the "station" (in this case, the airport), or even on the "train" (in this case, the plane). Second, a passenger dissatisfied with the accommodations (such as crying children in coach) can simply move to first class and pay the fare, if there is a seat available. Furthermore, following the high-speed rail business model would also reduce complexity as it would greatly reduce one of the airlines' biggest hurdles: baggage. In this scenario, passengers would be responsible for their own luggage, taking the burden off the airlines.

Returning to the role of enabling technology, it can also help an airline to develop business strategies to meet its (the airline's) own needs. For example, how can an airline implement some business strategies to "shockproof" itself from dramatic unexpected changes relating to, for example, significant fluctuations in the price of fuel and changes in the economy? Again, enabling technology can be leveraged to help develop some meaningful business strategies to maintain viability. Consider an example from outside the airline industry of a "self-destruction" strategy, in other words, doing it within your company first, so that the company can be ahead of its competitors. This business strategy of self-destruction is exemplified by the movie rental service, Netflix. Although the company started out as a mail order DVD rental company, it has been reported that it already had the vision right from the start that one day that the Internet could override, if not even replace, mail-order movie rentals, and therefore even named it "Netflix," instead of "Mailflix" or something of the sort.[14] Today, the company offers movies streaming from the Netflix.com site to a user's personal computer. It also offers streaming to TV-connected devices such as the Nintendo Wii, Microsoft Xbox 360, Sony PlayStation 3, or even a Blu-ray disc player. However, consumers may also still subscribe to the mail subscription service as well. Unlike most companies that either do not change or change too late (such as its struggling competitor Blockbuster), Netflix had the vision to see what could potentially destroy its business, and proactively take action accordingly.[15] Netflix utilized the aspect of business intelligence and analysis to stay ahead of the competition by predicting what could be coming ahead for its industry.

Within the airline industry, JetBlue is one example of an airline that has leveraged technology not only to develop, but also to continually adapt, its business strategy. Unlike many other businesses (both airline and non-airline), JetBlue appears to place technology at the core of its endeavors. One technology expert recognizes that the carrier actually leverages technology to differentiate itself and create a competitive advantage, unlike some competitors who tap into technology to provide in-flight offerings. JetBlue recognizes that the airline industry is highly prone to consolidation. However, rather than joining those who wish to pursue this path in an attempt to be the biggest, the carrier appears to believe that it (a) provides a superior offering (leather seats, ample legroom, movies, radio, satellite television), and (b) fills a niche (no full meal service, no lounges, one class of service, but operates more than one type of aircraft thus allowing it to serve smaller markets with smaller aircraft), and wishes to continue to do so. Therefore, the airline appears to have chosen proactively to upgrade its technology that enables it to easily form alliances with others, recognizing that the industry is inundated with capacity, and the choices for others will be to merge or to partner with airlines such as JetBlue. Specifically, JetBlue has invested in new technology that has enabled (a) the ability to determine product and service offerings at a very finite level (such as the ability to price features such as legroom differently for different markets), and (b) the flexibility to easily partner with others. The new technology has also yielded positive outcomes for travelers, as it gives them the ability to choose different features and services in alignment with their preferences. At the same time, the new technology has allowed JetBlue to capture this passenger data that, in turn, can be used to create personalized offerings for returning travelers. As one technology expert notes, for a business to remain sustainable, it must align and integrate its technology strategy with its business strategy.[16]

Takeaways

The major hurdle in the implementation of new-generation technology is the lack of clarity with respect to the identification of three strategies (business, information, and technology) and their alignment to identify and implement the next point(s) of integration where money can be made. This chapter illustrates one approach for sharpening the fuzziness of an airline's activities. First, there should a separation between the "I" and the "T" in information technology. Second, within the framework of the diamond, shown in Figure 6.1, information can be used to determine the next points of integration that add value and generate profit. Next, key elements of points of integration, namely people and processes, can be used to help determine the technology strategy. Finally, all three strategies (information, technology, and business) must be aligned on an ongoing basis to identify and implement new points of integration. The alignment of the four points of the diamond can be achieved by selecting an appropriate CIO, including the CIO in the strategy process from a holistic point of view, and ensuring that the interaction between the CIO and the CEO is organic and in real time. Airlines can gain significant insights from the experience of other businesses where technology has successfully enabled the identification and implementation of information and business strategies.

Notes

1   Peter Weill and Jeanne W. Ross, *IT Savvy: What Top Executives Must Know to Go from Pain to Gain* (Boston, MA: Harvard Business Press, 2009), pp. 5–6.

2   Peter Weill and Jeanne W. Ross, *IT Savvy: What Top Executives Must Know to Go from Pain to Gain* (Boston, MA: Harvard Business Press, 2009), pp. 6–7.

3   Case Study: "Isle of Capri Casinos Makes a Sure Bet," *Teradata Magazine*, Q1/2011.

4   "Border Management: Keeping Passengers Moving," *New Frontiers Paper*, SITA, 2010.

5   Michael Hammer and Lisa W. Hershman, *Faster Cheaper Better* (New York, NY: Crown Publishing, 2010), pp. 231–40.

6   Bill Tobey, "Data in the Driver's Seat," *Teradata Magazine*, Q2/2010.

7   Mercedes-Benz Press Release, September 2, 2010.

8   Peter Weill and Jeanne W. Ross, *IT Savvy: What Top Executives Must Know to Go from Pain to Gain* (Boston, MA: Harvard Business Press, 2009), pp. 29–32.

9   Victoria Moores, "Taking the Airlines Out of the Loop," *Airline Business*, March 2011, p. 32.

10 Christopher J. Varadi, "It's that easy with flydubai," *Airways*, April 2011, p. 46.

11 Ken Auletta, *Googled: The End of the World as We Know It* (New York, NY: The Penguin Press, 2009), p.7.

12 Ken Auletta, *Googled: The End of the World as We Know It* (New York, NY: The Penguin Press, 2009), pp. 7–9.

13 www.minuteclinic.com/

14 Damon Darlin, "Always Pushing Beyond The Envelope," *The New York Times*, August 8, 2010, p. 5.

15 Damon Darlin, "Always Pushing Beyond The Envelope," *The New York Times*, August 8, 2010, p. 5.

16 Debra Jacobs, Garrett Sheridan and Juan Pablo González, *Shockproof: How to Hardwire Your Business for Lasting Success* (Hoboken, NJ: John Wiley, 2011), pp. 72–4.

Chapter 7
Preparing for Competitive Renewal Opportunities

Airlines have performed very well in times of regular operations and reasonably well during modest irregular operations in light of their historic, but appropriate, focus on operations. The airline business is extremely complex given the nature of aircraft scheduling, crew rotations, maintenance requirements, airport slots, environment considerations, and weather conditions. In addition to airlines' operations centricity, passengers have also benefited significantly from the airlines' product centricity—hub-and-spoke systems, frequent flyer programs, in-flight entertainment systems, and lie-flat beds in premium cabins of long-haul, intercontinental flights. And now, airlines have started to progress toward customer centricity, exemplified by online check-in, dynamic packaging, mobile check-in, and, more recently, the beginning of online disruption management.

However, while many airlines have been moving from being operations-centric to product-centric to now becoming customer-centric, the pace of change has been slow given the rate at which passenger expectations have been changing. Passengers now expect to be in control 24/7, even while mobile, be recognized consistently through the entire travel cycle, and receive personalized service consistently during normal as well as irregular operations and across the travel cycle. At the same time, when more powerful and more frequent disruptions arise (with causes ranging from Icelandic eruptions to major blizzards to terrorist-related activities), older-generation business models have come under pressure to and have not been able to meet the needs of either passengers or airline businesses. Sometimes the information has been less timely. Other times the information has not been available across the entire travel cycle. The challenge has often related to the inflexibility of the technology systems.

The older-generation business models also worked reasonably well when pressures to reduce costs were relatively modest, there was plenty of low-hanging fruit to be picked on the cost side, and competition from low cost carriers and new full service global carriers was limited. Now, with increasingly discouraged operational workers (from pressures to continue to reduce operating costs) and with fares jumping around (due to fluctuations in the price of fuel, strategies to increase ancillary revenues, and the expansion of low cost carriers) customer satisfaction is declining for some segments of the traveling population, placing further pressure on the older-generation business models. On top of all these developments, competition is increasing at unprecedented rates. On one front are low cost carriers expanding in every region of the world and changing their business models to start moving toward the business model of legacy airlines. Some have begun to serve conventional airports, code-share, enter intercontinental markets, and merge with other low cost carriers. As mentioned in Chapter 1 some low cost carriers have begun to interline and code-share with global conventional network carriers. On a second front the Gulf-based carriers in the Middle East are continuing to expand their global networks and are offering superior service with equal or lower prices, exemplified by Emirates (selected by *Air Transport World* as the Airline of the Year for 2010). Some European passengers traveling to destinations in Asia, for example, find it more convenient to make transfers in the Persian Gulf than at major airports in their own countries—Charles de Gaulle, Frankfurt, and Heathrow, for instance. On a third front, are a few older-generation flag carriers who have begun to transform themselves from being conventional stereotypes to smart global players through the strategic deployment of their locations, synergistic working relationships with their governments, and visionary leaderships to become global players. Turkish Airlines would be one example that has already become the fourth largest airline in Europe with a significant growth in the number of passengers making connections at Istanbul's Ataturk Airport. On a fourth front, is the dramatic growth of the three major airlines in

China—Air China, China Eastern, and China Southern. These three airlines posted over US\$2 billion dollars in profit in the year 2010.[1]

To deal with these changes many legacy carriers have done well in reducing their costs through a wide array of methods such as optimization of fleet, networks, and schedules as well as through the consolidation process (M&A and alliances). They have also done well in the enhancement of ancillary revenues. And progress has been made in the area of customer experience—online and mobile check-in, for example. Now, continuing advancements in information and technology can enable airlines, particularly some large legacy carriers, to meet rising passenger expectations—recognition and consistent service delivery through the travel cycle—by becoming solution providers, aggregators of value, and managers of complexity.

Consider such airlines as American in the US, British Airways in Europe, and Japan Airlines in Asia. American must now compete with extremely large and powerful consolidated airlines such as United and Delta on one front, and large, well-branded, highly-productive, and fast-growing airlines on a second front (Southwest, JetBlue, and Virgin America). Ironically, many passengers consider that some US low cost carriers, such as JetBlue and Virgin America, actually provide higher quality of service within domestic markets than the conventional network carriers. In Europe, British Airways is not only facing stiff competition from its traditional, now consolidated, competitors (Air France and Lufthansa) on one front, increasingly powerful lower cost carriers (easyJet, Ryanair, and flybe) on a second front, but also the well-financed Gulf-based carriers in the Middle East (Emirates, Etihad, and Qatar Airways) with the enormous capacity to grow on the third front. Japan Airlines must now compete against not only All Nippon (strongly integrated in the Star Alliance), low cost carriers (such as Skymark and Air Do), the fast train network within Japan, but also rapidly growing airlines in Korea. Korean Air now connects multiple cities in Japan with multiple cities in China through Incheon International Airport. Yet, these conventional global airlines have powerful and, in some ways, insufficiently tapped resources (incredible networks, extraordinarily strong hubs, a wide variety of airplanes, enormous databases relating to frequent flyer programs—Lufthansa's Miles & More program has 20 million members,[2] and incredibly potent alliances to increase the scope of individual airlines). These *relatively* untapped resources, coupled with a strategic deployment of new-generation information and new-generation technology (used innovatively) can enable these older-generation conventional airlines around the world to become resilient and aggressive competitors not only for airlines with new business models (low cost, hybrid network, lower price and higher levels of service in intercontinental markets, and so forth), but also for well-healed and well-branded technology companies hovering around the distribution arena (Google, Apple, Amazon, and so on).

In light of the fundamentally and structurally changing marketplace, even newer-generation business models are needed by all categories of airlines that build upon the successes resulting from the optimization initiatives (related to restructuring, consolidation, simplification, and so forth). It is the innovation element in the new business models (facilitated by the deployment of information and technology) that can enable airlines to meet the contemporary needs of both passengers (hassle-free travel and personalized service, for example) and the airlines themselves (a reasonable return on investment through the normal business cycles and the ability to compete within and outside the airline industry, for example). How can such newer-generation information and technology help? This concluding chapter maintains that while information and technology on its own cannot solve the problems faced by passengers and airlines, management can go a long way in introducing innovation, depending on how they view:

- the next points of integration that add value
- the value of relevant and timely information on customers and their own operations to provide information-based customization
- the role of new-generation technology to gather the relevant information, mine it, and use advanced analytics to predict purchase behavior on the part of customers and operational behavior within different parts of the system

- the role of information and technology to offer a higher quality product (such as IFE, on-board connectivity) in an effort to become more customer-centric
- the value of competitive advantage to make the necessary investments in enabling information and technology to implement cost-effective business strategies
- the capability to bring about a change in corporate culture to deploy technology-enabled (and possibly even technology-driven) information and business strategies to develop and provide new transportation solutions

Let us first start with one view of a challenge and an opportunity. If one were to conduct a survey, there are still passengers and employees who feel that irregular operations are now occurring more frequently to the point that some passengers and employees consider them to be the new normal. On the part of passengers they feel that they are not being provided with adequate solutions to their problems during times of irregular operations. On their part while airlines have made significant progress in this area, especially in terms of providing increasing levels of individualized solutions for increasing levels of frequent-flier status, these types of initiatives need to filter down to infrequent travelers with lower-tier status or to even infrequent travelers with no status. For example, global, brand-named airlines do have processes in place to make personal calls (a "live" person instead of a computer) to their very top-tier passengers when a flight is cancelled and have an interactive conversation about options. The majority of passengers, however, simply get a message that their flight has been cancelled and that they have been re-accommodated on the next available flight. How acceptable is the re-accommodation provided by the airline is not usually discussed. What about other options such as booking the passenger on a competitor's flight that might involve a higher fare? All options can now be presented to passengers allowing each passenger to select the best price-service option for him or her—an opportunity to provide personalized service. It is the existence of such scenarios that will keep passengers patronizing the re-energized legacy carriers. Some technology experts are even suggesting that such options can be presented to passengers even on self-service check-in machines. However, it should be noted that in the case of *major* irregular operations (such as the volcanic ash incident in 2010), kiosks for check-in, bag retrieval, and Web check-in do not work, because airlines switch them off due to complexity issues. The consequence is that operations fall back to manned counters with long lines, hours of waiting time, creating huge costs and the highest degrees of customer dissatisfaction. Nonetheless, information and technology could be leveraged in more routine irregular operations to start moving toward offering more personalized solutions to the next levels/ tiers of passengers.

Consider another view of a challenge and an opportunity, the need to further grow ancillary revenues. Low cost airlines worldwide and US-based full service airlines have already made enormous strides in generating ancillary revenue through a wide array of sources. While the unbundling strategy is being evaluated and being adopted in other parts of the world (starting with alliance partners), the US carriers are exploring new sources (beyond charges for choice seats, priority boarding, and checking baggage) such as pre-ordered and customized food in the main cabin on intercontinental flights. Passengers get to choose the meal based on their willingness and ability to pay while airlines reduce their waste and get around the brand image of providing tasteless mass-produced food (and the pressure to reduce costs even further). The question is not if the airline products or services should be unbundled further, but rather, how various parts of the service can be unbundled, monetized, possibly re-bundled, and sold for the value created and value added without diluting the brand or creating negative reactions from passengers. Such a business strategy would be highly dependent on information that, in turn, would depend on an individual and vary by situation even for the same individual. Moreover, while third-party ancillary services are not new to the travel industry, they could prove to be a key driver in the future. Travel suppliers may leverage third-party ancillary services to yield such outcomes as the creation of brand differentiation through a higher level of personalization and the improvement of the travel experience across the travel cycle. One key difference between à la carte ancillary services and third-party services is that travelers may view à la carte ancillary services and their associated fees in a negative light as they may be services that were once included in the base price.[3]

In both cases business strategies can be developed that are much more customer-centric than operations-centric and product-centric, making airlines much more solution providers and value aggregators to provide personalized solutions for passengers by using information and technology much more innovatively. New business models can be developed, enabled, and, in some cases, even driven, by the deployment of strategic information and new-generation "listening to customer" technology. First, a customer-centric airline would recognize a customer from an integrated 360-degree viewpoint. Who is this customer, what is the value of this customer, what are the needs and habits of this customer across the travel cycle (at the website, at the airport, in the lounge, in the aircraft, and so on)? The next step would be to research and to offer customized solutions to all passengers (based on segmentation and willingness to pay) during a broad spectrum of regular and irregular operations. This process would appear to result in enormous manpower costs, a change to numerous systems so as to be able to extract information from traditional sources (such as schedules and availability of all airlines) as well as non-traditional sources such as schedules and availability of alternative modes of transportation, relevant for the situation. There is also the issue of standards in coordinating the schedules of other modes of transportation.

However, this is where new-generation technology comes into play. Given the feasibility of every passenger having a unique digital identity, the availability of ubiquitous low cost communications and cloud computing, the convergence of various devices, the emergence of enormously powerful search engines, the enormous and low cost processing power of new-generation systems, all passengers can be contacted virtually 24/7 and offered a much richer array of options on a personal basis— information-based mass customization. For example, an airline could not only inform a passenger about the availability of other modes of transportation but, in fact, offer to make reservations on buses, trains, and limousines and provide written confirmations through a broad spectrum of channels. The bottom line is that timely, relevant, and actionable solutions can be offered by an airline to meet its passengers' individual needs. Consider business passengers who travel on a fairly regular basis but who have not achieved top-tier status. It is the needs of this segment that the airlines can start to address. Returning to the scenario discussed throughout the book, automatically rebooking a passenger on the next evening's flight across the Atlantic when his flight for this evening is cancelled is timely and actionable, but is hardly relevant when the passenger has a commitment first thing in the morning after the arrival of the initially booked flight. If the passenger has to go and redo his reservation anyway since the rebooked flight does not meet his needs, then the process is not efficient. If the issue is one of higher costs to provide options of solutions then one option could be to make this service available on a per charge basis.

Just as there is a fee for a better seat or an extra piece of luggage, there could be a fee for travel concierge service on a 24/7 basis, such as the three-tier service suggested in Chapter 3. For the passenger it is the ability to buy a customized solution to an unforeseen problem. For the airline it is a way to improve the passenger's travel experience by reducing stress and at the same time generate ancillary revenue through a source that passengers would be much more willing to pay, although at different prices for different levels of service provided. The concept of travel concierges is not new. It is already being offered. There are third parties that offer destination guides for travelers who might need to locate, for example, hotel accommodations, restaurant guides and reservations, and tickets for events. The concept that has already proven successful for some leisure travelers can now be expanded, given the proliferation of the mobile channel and the virtually unlimited low cost bandwidth to provide rich solutions during irregular operations. If large, resource-rich, and well-established airlines were to provide such services, they could easily compete with the lower cost, new-generation airlines even with their higher unit costs. For their part, if the new-generation airlines did not have such resources or did not want to develop and offer such ancillary services for fees, they could offer to work with branded third parties to develop and provide such services and charge commissions for the sale of such services to passengers.

One other benefit of selling the services of third parties is that passengers would not view these services as something included previously in the bundled product.

The critical success factor for innovating business models (based on information and technology) is the ability for management to develop a "foresight" through developing an "insight" into the role of information and the use of enabling technology to align information and business strategies. Keep in mind that information is the core of the Internet and mobility is the heart and soul of travel. The combination of the two can be very powerful in providing travel solutions, especially in terms of location-based services. Consider the following scenario. A passenger calls a low cost airline and asks the price of a ticket from A to B the next morning. The airline quotes a low fare but informs the traveler that there are no seats on the flight. Then the passenger calls a major information and technology-driven airline (along the lines discussed) and is informed that there is always a seat available on that airline if the passenger is willing to pay for it. The airline can develop an ability to call any number of the passengers who are holding confirmed reservations and provide various levels of incentives to free up a seat. Under such circumstances there are passengers who will simply start calling the latter airline knowing that there is always a seat available for them regardless of the destination and the date of travel. The issue is simply one of a price. The information and technology-armed airline can provide a wide spectrum of price-service options.

Consequently, it is management that must first develop an aptitude for a fundamental change (to evolve from operations and product centricity into true customer centricity) to deploy information and technology to manage complexity to become solution providers and integrators of value. The issue is no longer whether technology is available to capture the right information on a personal basis to enable an airline to develop a new business model; rather, it is related to the area of focus of an airline's customer service strategy. Some airlines are focusing on innovation in the cabin. For its first-class passengers on some long-haul flights Emirates provides private suites (with sliding doors), on-board shower facilities, and a more than 1,000-channel in-flight entertainment system. Alaska Airlines has chosen to use technology to make airport processing more efficient (as described in Chapter 3). By replacing traditional check-in counters with multiple kiosks and bag-check stations, the average check-in time at Seattle Airport has been reported to be less than two minutes (even at peak times) compared to about 30 minutes using the standard check-in process.[4] Alaska Airlines is really striving to be understanding of travelers' time constraints. Take its recent decision regarding its baggage guarantee. If a passenger's luggage is not at the carousel within 20 minutes of the plane arriving at the gate, the passenger receives her choice of a financial compensation (US$20) toward future travel or mileage compensation. Then there is Air New Zealand that is using technology both in the cabin and on the ground to improve passenger experience—the next level of customer centricity. In the cabin, the airline has achieved significant innovation in the design of a lie-flat seat in the economy class by converting a set of three normal economy-class seats into a relatively flat seat. See the sequence of photos in Chapter 6. On the ground, passengers traveling within domestic markets are no longer required to check in if they already have selected a seat and obtained a boarding pass prior to arriving at an airport. Those traveling without a bag can proceed directly to the gate upon arrival at an airport.

While information and technology are now available for airlines to introduce greater innovation in their business models, each airline must decide on an optimal balance between its own needs and the needs of its customers. Take, for example, the current focus of most airlines to maximize the generation of ancillary revenue. An airline could clutter up its website with all kinds of sale offers. Likewise, it could offer an incredible number of offers at the check-in kiosks and sell lounge passes. However, such strategies could easily turn some passengers away from the website of an airline, lengthen the time to check-in at the kiosks for passengers who simply want to check-in and not wish to purchase additional services, and reduce the positive experience of visiting an airport lounge. The ancillary revenue strategy must therefore be aligned with the brand strategy. Should an airline be still promoting itself as a full service airline if the product has been totally unbundled with a price tag for every component?

Passengers are looking for recognition across the entire travel cycle, services that are personalized (relevant, timely, and meet their needs) and delivered consistently across channels and the travel cycle (by every employee and at every touch point). Passengers also want more control, more choice, and an enhanced travel experience during regular and irregular operations. As highlighted throughout this book, information and technology are the building blocks for this customer centricity within the mass customization framework. It is information-driven business strategies, enabled by technology that can permit potentially powerful legacy airlines to become solution providers and integrators of value (through alliances with other airlines and related service providers). Despite their higher operating costs they can become airlines of choice. Innovative use of information and technology, along with the sophisticated use of analytics, can now be used to benefit all segments of the traveling public, based on services desired and willingness to pay. This book closes with some calls for action by airline managements:

- Airline CEOs must continue to innovate their business models with much more focus on not just immediate and traditional competition, but opportunities and challenges at the edges of the business and at the edges of the market.[5] Senior management must free up time from exploring the historical ways of reducing costs and growing revenue to envision more creative ways to identify and implement flawlessly information-based and technology-enabled strategies.
- CEOs must continue to fine-tune the organizational structure and corporate culture to *listen* more effectively to passengers in real time and from a broad spectrum of sources deploying new technology to deliver value by determining the next points of integration.
- CEOs must demand the availability of comprehensive and timely business intelligence, relating not only to customers, but also to the airline's assets and people.
- As for CIOs, they should take on the responsibility for some of the non-alignment of business and enabling technologies and overlooked integration opportunities, the results of

failure to deliver adequate business intelligence capabilities, and adequate actionable metrics. Without the appropriate business intelligence or metrics an airline is likely to struggle in terms of setting strategy and finding new opportunities.[6]

- CEOs, in turn, must sharpen their focus in dealing with the hurdles mentioned in Chapter 5. The CEOs must give the CIOs some strategic focal points, so that they can develop the technology organization with the smarts that are required.

In the final analysis, one seasoned observer, with experience in both the technology and the airline business, noted: consider a world-class skier. In addition to the traditional strengths of the athlete, it is the technology/equipment and advanced processes in practicing that are required to compete on a world-class level. An athlete could hardly win an Olympic race these days using the same pair of skis used in a race more than ten years ago. Similarly, airlines cannot be successful in tomorrow's marketplace using antiquated information and technology systems as well as processes, or piecemeal approaches. Information and technology must be deployed strategically and holistically and implemented flawlessly.

Yet, adopting the latest available technology is not enough. The step-changing "me-centric" passenger shopping/buying behavior (enabled by powerful search engines, mobile devices, and applications) is a game changer for the airline's merchandizing model. Passengers now want to customize their travel experience by managing their own travel with respect to what they buy, when they buy, how they buy, and where they buy. And they only want to pay for the products and services they value. Such a buyer/seller merchandizing scenario is not new to some other businesses with a focus on customer centricity motivated by customer-driven innovation. Airlines must transform their business models to dynamically target passengers and their valued products and services. As one executive stated, passengers are looking for new "buying systems," not new "distribution systems." Consequently, airlines must identify and implement innovative ways to support the "sourcing" focus of the "me-centric" customers' systems instead of relying on advancements on the "distribution" capabilities of their own systems.

Notes

1   Daniel Tsang, "Reversal of fortunes," *Orient Aviation*, Vol. 18, No. 2, March 2011, pp. 24–6.
2   "Miles & More record: 20 million members," *Air Transport News*, 11/02/2011.
3   "Cross-Sell Your Way To Profit," Forrester Consulting, January 2011.
4   "ATW's 37th Annual Airline Industry Achievement Awards," *Air Transport World*, February 2011, p. 32.
5   Paul Nunes and Tim Breene, "Reinvent Your Business Before It's Too Late," *Harvard Business Review*, January–February 2011, pp. 80–7.
6   For excellent insights on the evolving role of CIOs in both public and private sectors, see "The Essential CIO: Insights from the Global Chief Information Officer Study," *CIO C-suite Studies*, IBM, May 2011.

Appendix
BSI: Teradata Webisode: The Case of the Mis-Connecting Passengers

Consider the following scenario developed by a "Business Scenario Investigations" (BSI) team. In this case, the researchers at BSI: Teradata have been hired to propose a complete overhaul of Air London's rebooking application to improve its customer satisfaction ratings. An online "webisode" of this case appears on YouTube at http://www.youtube.com/watch?v=NXEL5F4_aKA.

Air London has a hub at Frankfurt Airport. As an example, there are many flights arriving at Frankfurt that are connecting to Air London's flight to London. Unfortunately, three have been delayed: one from Rome, one from Tokyo, and one from Johannesburg. There are four impacted passengers (the Rome flight has two passengers), who will roll over to the next (and last) flight of the day for Air London. Because there are only two available seats, two passengers must stay overnight. Two can go tonight, and two must wait until tomorrow morning, but even that first flight has only two seats.

Currently, Air London Reservations uses typical rules to determine which two passengers would be chosen. These rules tend to be based on such criteria as frequent flier program tier status, the price of the ticket, and the point of origin.

The BSI team helps Air London by asking: what other factors could be used to improve customer satisfaction ratings? What data could be used for a new rules-based engine that includes more real-time factors and input from each of the four passengers? The team starts by examining passenger information for the Frankfurt to London mis-connecting passengers. The airline uses cameras at check-in to capture photos of the passengers. Following are the photos and profiles of the impacted passengers, shown in Figure A1 and A2, respectively:

- Passenger 1, Jason Chang, is flying from Cairo to Rome to Frankfurt and then connecting to London. He is flying on a cheap ticket and has 4,522 frequent flyer miles this year, with a total of 128,750 lifetime miles with the carrier.
- Passenger 2, Lana Kaliki, is coming in on the same flight from Rome. She has 20,250 lifetime miles, 4,100 this year, and is on an average priced-ticket.
- There is no background information regarding Passenger 3, Steffi Schneider, arriving on the Tokyo flight, indicating that she is a first-time flyer with Air London. She paid full fare, 4,280 euros, for her ticket from Narita.
- Passenger 4, Conrad Hunter, is a tier-one customer with 84,000 miles already this year, and 261,500 lifetime miles with the carrier.

Figure A1   Four Impacted Passengers Missing LHR Flight

Source: Teradata and Wilson Advertising (printed with permission)

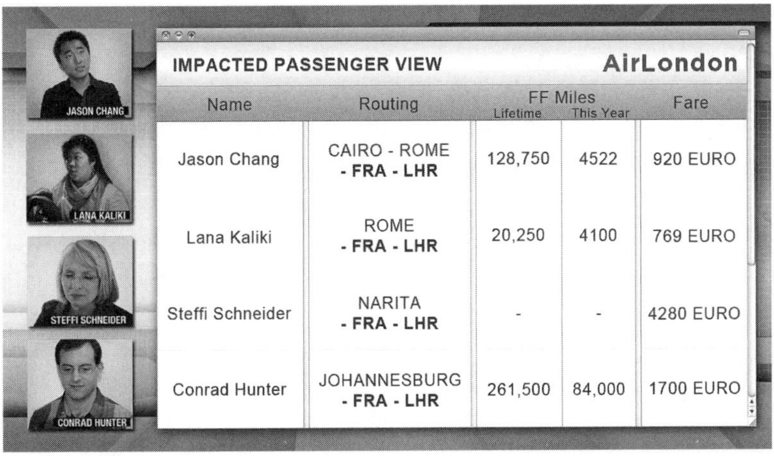

Name	Routing	FF Miles Lifetime	FF Miles This Year	Fare
IMPACTED PASSENGER VIEW				**AirLondon**
Jason Chang	CAIRO - ROME - **FRA - LHR**	128,750	4522	920 EURO
Lana Kaliki	ROME - **FRA - LHR**	20,250	4100	769 EURO
Steffi Schneider	NARITA - **FRA - LHR**	-	-	4280 EURO
Conrad Hunter	JOHANNESBURG - **FRA - LHR**	261,500	84,000	1700 EURO

Figure A2   Impacted Passengers—Basic Information

Source: Teradata and Wilson Advertising (printed with permission)

A simple implementation would make decisions based only on this data. But financial information can be added to the rebooking system to create deeper customer insights. This information can include financial contribution scores, including lifetime value predictions as well as current year revenues. It can also include frequency of bookings and profit margins. Looking in the database (as depicted in Figure A3), Reservations can see that Jason only books the lowest margin flights. Lana books at the last minute, but pays full fare so she is a very high margin customer. She is also booking more frequently. Steffi only has current year information since she's new to AirLondon. For Conrad, it looks like his company's booking engine is forcing lower-margin choices for him. Based on this additional financial information, who should be chosen to go to London? This information might lead to a different conclusion regarding the two passengers to be given this evening's reservations.

However, before making the decision, BSI investigators think AirLondon might want to consider even more factors (see Figure A4). Further information can be added such as which booking channels travelers use and their associated costs. For example, Jason books on the Web, an inexpensive channel. We can also see from the data that Jason checks out competitor websites, indicating price sensitivity. Lana only books through the call center, never online. While such contacts cost more to serve, additional information can be obtained by analyzing call-center logs and it's noted that on the last call, she asked about fares for flying 20 extended-family members from London to Chicago for a family reunion. Steffi appears to have booked through a travel agent, while Conrad always uses his corporate Web engine to book, so that is low cost to the carrier. These new channel booking costs and uses might drive yet a different decision about who goes to London tonight.

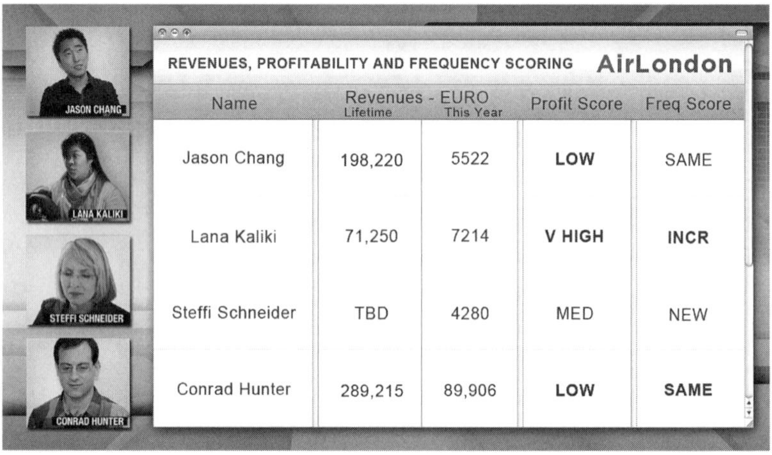

Name	Revenues - EURO Lifetime	This Year	Profit Score	Freq Score
Jason Chang	198,220	5522	LOW	SAME
Lana Kaliki	71,250	7214	V HIGH	INCR
Steffi Schneider	TBD	4280	MED	NEW
Conrad Hunter	289,215	89,906	LOW	SAME

Figure A3 Revenues, Profitability and Frequency Scoring
Source: Teradata and Wilson Advertising (printed with permission)

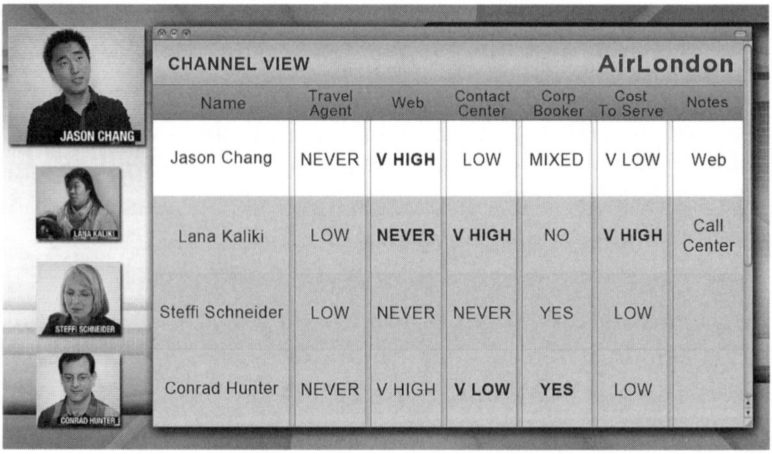

Name	Travel Agent	Web	Contact Center	Corp Booker	Cost To Serve	Notes
Jason Chang	NEVER	V HIGH	LOW	MIXED	V LOW	Web
Lana Kaliki	LOW	NEVER	V HIGH	NO	V HIGH	Call Center
Steffi Schneider	LOW	NEVER	NEVER	YES	LOW	
Conrad Hunter	NEVER	V HIGH	V LOW	YES	LOW	

Figure A4 Channel Use
Source: Teradata and Wilson Advertising (printed with permission)

In addition to using all the data within the rebooking engine, Air London also could build a better customer-centric display that frontline staff such as gate agents can use to give them a bird's-eye view of what is occurring in terms of each individual passenger (including the value of a customer and any real-time status). For example, Figure A5 shows four different portlets. Such operational real-time information helps the agents when talking to each customer. In this case, some additional information on the "Current Flight" screen includes the fact that Lana checked in carrying an infant and that the baggage from the Rome flight did not make the Frankfurt flight, impacting both Jason and Lana. Information from the call-center notes (on Steffi's screen, not shown) indicates that Steffi's original flight from Tokyo was cancelled, and that she waited six hours for her next flight. Steffi used the time to call the care center and she does not appear to be happy given all the cancellations and delays, which the agent can see via call-center text on the screen.

As usual, other dynamic factors may impact decision-making on the fly. In this case, the Web booking engine did not get turned off while rebooking was in progress, and one of the two remaining seats was sold on the next morning's flight. Therefore, now the first flight tomorrow morning only has one free seat, instead of two. The second flight at noon has a dozen seats available. The plot thickens!

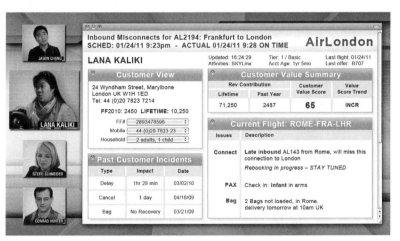

Figure A5   Screen for Passenger Agents—LANA

Source: Teradata and Wilson Advertising (printed with permission)

This exercise illustrates that there are many factors that Air London can use to make better rebooking decisions, leading to an improvement in customer satisfaction. The good news is most of the information is readily accessible; it is a matter of loading it into an active data warehouse so that it is all in one place, real-time analytics can be performed, and information can be made widely available to frontline employees and systems at the points of customer contact.

One future opportunity proposed by the BSI consulting team is that that the carrier should plan to better use in-flight Internet seatback capabilities. The carrier could use pop-up screens to interact with impacted customers while they are still in flight, giving them options and gathering more information about what they want to do when there is a mis-connect situation. The system could interact with them in priority order based on the rebooking score rankings. Suppose the tentative ranking was: Conrad, Jason, Lana, Steffi. The system would begin lighting up screens one by one to interact with customers on their rebookings while in the air.

Figure A6 shows what Conrad might see and proposes some options. His choices could impact the others.

Figure A6   Creative Idea — Initiate Interaction with an Individual Passenger

Source: Teradata and Wilson Advertising (printed with permission)

It turns out that Conrad is happy to stay in Frankfurt for an evening after his long flight from Johannesburg to rest, and then take the flight to London around noon the next day. The system rebooks him automatically and makes reservations for a hotel and dinner, as shown in Figure A7.

Figure A7   Creative Idea—Close Interaction with an Individual Passenger

Source: Teradata and Wilson Advertising (printed with permission)

The system then interacts with Jason, who lives in Frankfurt, and if he is going to miss his flight to London, he will miss a meeting he did not want to attend anyway. So he is fine with canceling the last leg of his trip. Thus, by having interactive capabilities, Lana and Steffi, who were originally the last choices, could actually be accommodated on the last flight.[1] Case solved! The point of the case is that leveraging real-time passenger interaction can yield a win-win situation, lower costs to the airline and higher customer satisfaction for passengers.

Notes

1   BSI: Teradata "The Case of the Mis-Connecting Passengers," January 2011 uses fictional companies and characters (script). Based on http://www.youtube.com/watch?v=NXEL5F4_aKA (video on YouTube) and conversations with Dr. David Schrader. Copyright © 2011 by Teradata Corporation All Rights Reserved. Produced in U.S.A.

Index

About the Author

Nawal Taneja's career in the global airline industry spans 40 years. As a practitioner, he has worked for and advised major airlines and airline-related businesses worldwide, facilitating their strategies. His experience also includes the presidency of a small airline that provided schedule and charter services with jet aircraft, and the presidency of a research organization that provided advisory services to the air transportation community worldwide. In academia, he has served as Professor and Chairman of Aerospace Engineering and Aviation Department at the Ohio State University, and an Associate Professor in the Department of Aeronautics and Astronautics of the Massachusetts Institute of Technology. On the government side, he has advised civil aviation authorities in public policy areas such as airline liberalization, air transportation bilateral and multilateral agreements, and the financing, management, and operations of government-supported airlines. He has also served on the board of both public and private organizations.

At the encouragement of and for practitioners in the global airline industry he has authored six other books:

- *Driving Airline Business Strategies through Emerging Technology* (2002)
- *AIRLINE SURVIVAL KIT: Breaking Out of the Zero Profit Game* (2003)
- *Simpli-Flying: Optimizing the Airline Business Model* (2004)
- *FASTEN YOUR SEATBELT: The Passenger is Flying the Plane* (2005)
- *Flying Ahead of the Airplane* (2008)
- *Looking Beyond the Runway: Airlines Innovating with Best Practices while Facing Realities* (2010).

All six books were published by the Ashgate Publishing Company in the UK.

He holds a Bachelor's degree in Aeronautical Engineering (First Class Honors) from the University of London, a Master's degree in Flight Transportation from MIT, a Master's degree in Business Administration from MIT's Sloan School of Management, and a Doctorate in Air Transportation from the University of London.